PHYSICS OF
SOLID SOLUTION
STRENGTHENING

The Metallurgical Society of AIME Proceedings
published by Plenum Press

A Publication of The Metallurgical Society of AIME

PHYSICS OF SOLID SOLUTION STRENGTHENING

Edited by

E. W. Collings
Battelle Columbus Laboratories

and

H. L. Gegel
Wright Patterson Air Force Base

PLENUM PRESS·NEW YORK AND LONDON

Library of Congress Cataloging in Publication Data

Main entry under title:

Physics of solid solution strengthening.

"Proceedings of the ASM-TMS(AIME) symposium held in Chicago, Illinois, October 2, 1973."
"A publication of the Metallurgical Society of AIME."
Includes bibliographical references and index.
1. Solution strengthening–Congresses. I. Collings, E. W. II. Gegel, H. L. III. American Society for Metals. IV. American Institute of Mining, Metallurgical and Petroleum Engineers.

TN689.2.P49	669'.94	75-33368

ISBN 978-1-4684-0759-4   ISBN 978-1-4684-0757-0 (eBook)
DOI 10.1007/978-1-4684-0757-0

Proceedings of the ASM–TMS(AIME) Symposium held in Chicago, Illinois, October 2, 1973

© 1975 Plenum Press, New York
Softcover reprint of the hardcover 1st edion 1975
A Division of Plenum Publishing Corporation
227 West 17th Street, New York, N.Y. 10011

United Kingdom edition published by Plenum Press, London
A Division of Plenum Publishing Company, Ltd.
Davis House (4th Floor), 8 Scrubs Lane, Harlesden, London, NW10 6SE, England

Preface

This book is the proceedings of a Symposium entitled "The Physics of Solid-Solution Strengthening in Alloys" which was held at McCormick Place, Chicago, on October 2, 1973, in association with a joint meeting of the American Society for Metals (ASM) and The Metallurgical Society (TMS) of the American Institute of Mining, Metallurgical, and Petroleum Engineers (AIME). The symposium, which was initiated and organized by the editors of this volume, was sponsored by the Committee on Alloy Phases, Institute of Metals Division, TMS, AIME, and the Flow and Fracture Section of the Materials Science Division, ASM.

The discipline of Alloy Design has been very active in recent years, during which considerable stress has been placed on the roles of crystallography and microstructure in the rationalization and prediction of properties. Underestimated as a component of alloy design, however, has been the importance of physical property studies, even though physical property measurements have traditionally been employed to augment direct or x-ray observations in the determination of phase equilibrium (and, indeed, metastable equilibrium) boundaries.

In addition to studies of crystallographic structure and microstructure which will continue to play dominant roles in alloy design and application, due regard must be paid to the atomic interaction as a basis not only for dislocation theory and theories of solution strengthening, but also for an understanding of the influences of trace elements on important mechanical properties such as high-temperature creep and stress-rupture; and on grain-boundary-controlled effects in general. Thus even though some "impurities" (either metallic, metalloid, or gaseous--referring

to their bulk physical forms) act as strengtheners when in solid
solution, they may have severe deleterious influences if present
in grain-boundary precipitates. Similarly, those same atomic
interactions which favor solution strengthening by a given solute
species when present in sufficient dilution, may lead to brittle-
ness in the high-concentration regime where intermetallic compounds
exist either as single phases or as second-phase precipitates.
With these thoughts in mind, the symposium was organized and,
eventually, this book was assembled.

In its planning stages a symposium was envisioned which dealt
with solution strengthening and related effects on various levels,
from atomic through macroscopic. Speakers were invited, each
eminent in his field, who would address themselves to the selected
subject areas. Aspects to be presented were atomic interactions;
dislocations; solute-dislocation interactions; computer simulation
of atomic motion; physical effects, considered from phenomenological,
atomic (or chemical), and electronic standpoints; influence of solutes on bulk (as distinct from local) properties--hence elastic
modulus effects; and finally, the common ground between atomic-
scale impurities, and macroscopic observables such as grain size
and grain boundaries. It is often considered useful to "keynote"
a symposium. However, a keynote address is usually too late to be
effective if delivered at the time of the conference--the other
speakers having already decided on their contributions. Thus, in
an attempt to confer some degree of cohesion to the prospective
meeting, a "keynote reprint" entitled "A Physical Basis for Solid-
Solution Strengthening and Phase Stability in Alloys of Titanium"
was enclosed with each invitation to submit an abstract.

In this work, the sequential arrangement of the papers, which
is slightly different from that in which they were presented at the
symposium, is in accordance with an atomic-to-macroscopic logic.
The book starts out with detailed discussions of point-defect-solute
interactions considered from the standpoints of physical metallurgy
and the "chemistry" of atomic interactions; to be followed by discussions of the interatomic potentials associated with dissolved
impurities. Using such potentials as starting points, it is possible to simulate the movement of atoms and dislocations, hence to
quantitatively describe the atomistics of diffusion, solution
strengthening, and fracture. Two papers deal with computer simulation of effects associated with point defects. The book continues
with descriptions, in terms of physical properties and thermodynamics, of macroscopic observables associated with solution strength-
ening--a treatment which is followed and augmented by a discussion
of alloy theory as it applies to that topic. As well as having
local or short-range influences, solutes may affect bulk properties,
especially if they are able to be present in relatively high concentrations. Accordingly, a paper is included which deals with

elastic effects associated with alloying, a natural sequel to which
is the penultimate paper which presents electronic models for the
calculation of several bulk alloy properties. Practical alloys are
not generally encountered in monocrystalline form. It is therefore
necessary to consider, in the context of solute-solvent interactions,
the effects of grain size and grain-boundary precipitation; and this
is done in the final paper.

It is hoped that this book will provide a useful review for
both the scientist and the student of this expanding area of
research, and that it will stimulate further interest in those
important aspects of alloy physics which underly solution strengthening in metals.

Acknowledgments

The sponsors of the technical work described in this book,
viz. the United States Air Force and the (then) United States
Atomic Energy Commission, are recognized, when appropriate, at the
end of each chapter.

The editors wish to acknowledge Mrs. Jean Gwinn, who typed a
first draft of this book, and the conscientious and devoted efforts
of Miss Ritamay Jennings who typed, and patiently corrected, the
entire set of manuscripts in camera-ready form; and who was responsible for the key-punching, layout, and typing associated with index
preparation.

E. W. Collings
Battelle
Columbus Laboratories
Columbus, Ohio, USA, 43201

H. L. Gegel
Air Force Materials Laboratory
Wright-Patterson Air Force Base
Ohio, USA, 45433

Contents

STRENGTHENING OF ALPHA TITANIUM BY THE

INTERSTITIAL SOLUTES C, N, AND O

H. Conrad, B. de Meester, M. Döner, and K. Okazaki

Metallurgical Engineering and Materials Science Department

University of Kentucky
Lexington, Kentucky 40506

ABSTRACT

The experimental data on the strengthening of α-Ti by the interstitial solutes C, N, and O are reviewed and compared with the two principal models for solid solution strengthening, namely the Fleischer-Friedel and the Mott-Nabarro models as recently modified by Labusch. Furthermore, the nature of the interactions between the moving dislocations and the interstitial solute atoms is considered. It is concluded that the strengthening due to the interstitial solutes for prism slip in single crystals and for the plastic flow of polycrystals is described reasonably well by an interaction on the first-order prism planes according to the Fleischer-Friedel model. The experimentally derived interaction energy is in accord with that calculated for the combined effect of the size misfit and the modulus mismatch or that for the breaking of chemical bonds between the interstitial solutes and the surrounding Ti atoms.

1. INTRODUCTION

It is now well established [1-12] that the interstitial solutes C, N, and O have a pronounced effect on the strength of α-Ti (hcp) at temperatures below about $0.5\ T_m$. At temperatures below about 600 K ($\sim 0.3\ T_m$), the effect has been attributed to the interaction between dislocations moving on the first order prism planes and stationary interstitial solute obstacles [4-9], whereas in the temperature range of 650 to 850 K (0.3–$0.45\ T_m$) it appears to be due to the interaction of moving dislocations with moving interstitial solute atoms, i.e., dynamic strain aging [10-12]. In this latter

1

temperature range, interstitial solutes principally influence the form of the stress-strain curve, strain hardening, and ductility; they have little, if any, effect on the level of the yield stress.

In the present paper, we will concern ourselves only with the strengthening due to the interstitial solutes C, N, and O in the low-temperature region, where these interstitial solutes are expected to be immobile. We will first review the experimental data on the subject; this will be followed by a consideration of the model of solid solution strengthening applicable to the data and the parameters which may be derived therefrom; finally, the dislocation-interstitial atom interaction(s) which may be responsible for the strengthening will be considered.

2. EXPERIMENTAL DATA

2.1 Material

Most of the experimental data on polycrystalline material covered in this review have been obtained by the present authors and their co-workers using tensile tests on wire specimens of about 1.6-mm dia. with grain sizes of 1 to 20 μm. The three-dimensional grain size distribution in these specimens is approximately log-normal and constant for a given average grain size, independent of the combination of the time and temperature of the annealing treatment [13,14]. The texture consists of a strong $<10\bar{1}0>$ wire texture for grain sizes of the order of 1-2 μm, which decreases in intensity as the grain size increases, and splits into two components (each approximately 15° from the wire axis) for grain sizes larger than about 5 μm [13, 15]. Although most data on Ti-interstitial alloys considered here have been on wire specimens of 1 to 20-μm grain size, results obtained on rod, plate, and sheet specimens of the same or larger grain sizes are in general accord with those from the wire specimens [5,16,17]. However, one should be alert to some difference in the magnitude of the stresses when there exists an appreciable difference in texture; such differences are very pronounced in sheet when the stress state is biaxial [18].

The base material used in preparing the Ti-interstitial alloys of Fig. 1, and the base for comparisons in general, is the 99.9+ MARZ grade, zone-refined Ti produced by the Materials Research Corporation, Orangeburg, New York. The nominal composition of this material provided by the supplier is given in Table I. However, chemical analyses by a number of independent laboratories yielded interstitial contents somewhat in excess of the nominal quoted by the supplier, with considerable scatter from laboratory to laboratory and lot to lot [13]. The reasons for these discrepancies are not clear. The single crystals of this material employed by the authors

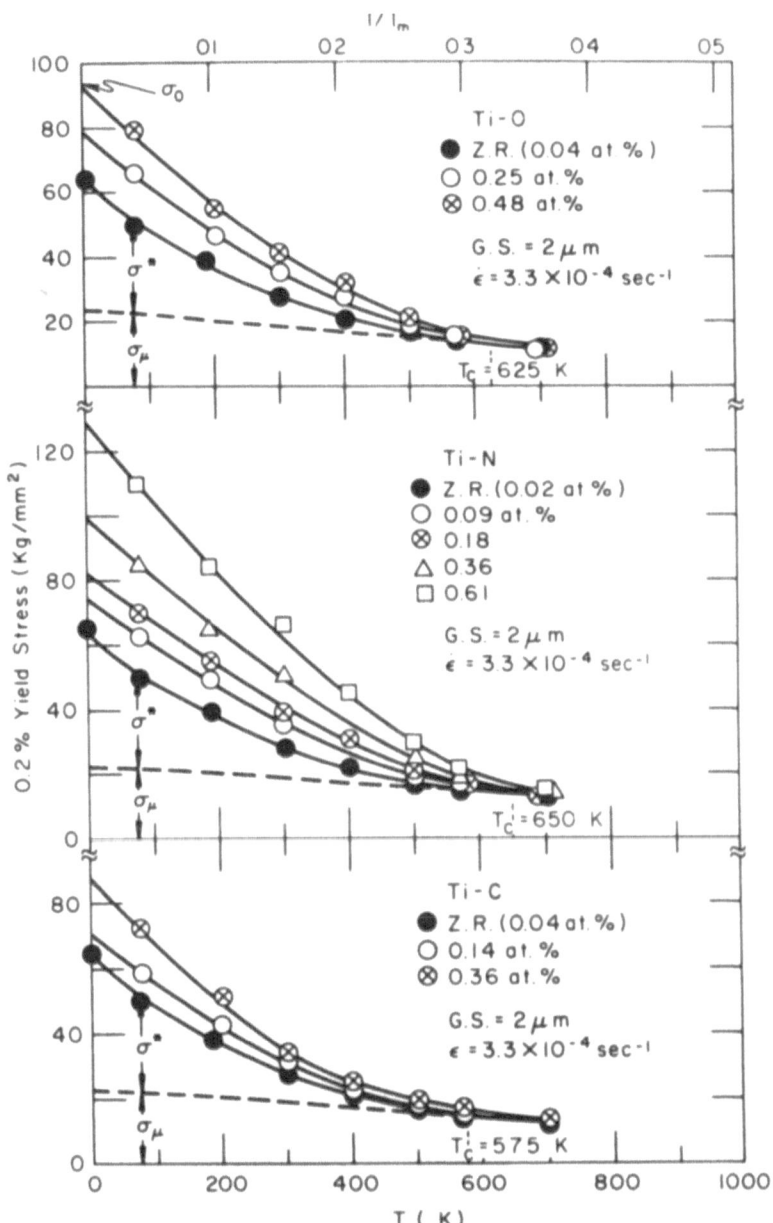

Fig. 1. Effect of a temperature on the 0.2% yield stress of poly-crystalline Ti as a function of interstitial content. Data from Refs. [7-9].

TABLE I. Chemical Composition of the MARZ Zone-refined Titanium in ppm by Weight
Titanium in ppm by Weight

Source	Ref.	Fe	Cr	Mn	Mg	Al	Si	C	N	O	H
MRC (Nominal)		50	12.0	1.2	<15	15.0	3.5	78	6	63	4
Okazaki and Conrad	13										
Max.								85	130	790	26
Min.								70	4	182	4
Avg.								78(2)	58(9)	424(11)	17(5)
Okazaki, et al.	7-9							160	103	133	---

Note: Numbers in parenthesis next to average composition give the number of determinations.

were generally 0.63-cm dia., while the polycrystalline specimens
were generally in the form of 0.16-cm dia.-wires as mentioned above.

2.2 Effects of Temperature, Strain Rate, and Interstitial Content on the Flow Stress

The effect of temperature on the tensile flow stress of the
polycrystalline Ti-interstitial solute alloys depicted in Fig. 1
is typical in that at low temperatures the flow stress σ is sensi-
tively dependent on the temperature, T (and strain rate, $\dot{\varepsilon}$), whereas
at higher temperatures it varies with temperature only as the shear
modulus, μ.[†] Following the usual procedure, one can then separate
the flow stress into two components

$$\sigma = \sigma^* (T, \dot{\varepsilon}) + \sigma_\mu \qquad (\sigma^* = 0; \ T \geq T_c). \tag{1}$$

It has been shown for Ti wire specimens of 1 to 20-μm grain size,
and strains of the order of 0.2%, that the value of σ_μ derived by
the back-extrapolation technique (illustrated in Fig. 1) is in good
accord with that obtained by other techniques [22-25]. Hence, it is
concluded that any change in the dislocation structure such as has
been observed following straining of Ti at various temperatures
[26-28] does not significantly influence the value of σ_μ at the 0.2%
yield stress. Support for this conclusion is that the values of σ^*
obtained from temperature-cycling, strain rate-cycling, and stress
relaxation during an otherwise constant strain rate tensile test
are in good accord with those obtained using the back-extrapolation
technique [23,29-32].

The results of Fig. 1, and other data [4-9], indicate that the
effect of the interstitial solute concentration C_i on the flow stress
of Ti is mainly on σ^*. On the other hand, the effects of strain ε
and grain size d are primarily on σ_μ [4-9,17]. Moreover, it was
found (see Fig. 2 and also Ref. [26]) that σ_μ is given by $\alpha\mu b\rho^{\frac{1}{2}}$,
where α is a constant of about 2.0, μ the shear modulus, b the

---

[†] As pointed out by Kocks [19], the appropriate modulus μ to use for
the interaction of dislocations with discrete obstacles is $\sqrt{K_s K_e}$.
For dislocations on the first-order prism planes in hcp crystals,
$K_s = (C_{44} C_{66})^{\frac{1}{2}}$ and $K_e = (C_{11}^2 - C_{12}^2)/2 \, C_{11}$ [20]. Hence, $\mu = \sqrt{K_s K_e}$
is used throughout this paper, employing the elastic constants re-
ported by Fisher and Renken [21]. In previous papers by the authors
[5-9], C_{66} (which is the most temperature-sensitive elastic constant)
was used for the shear modulus. The use of $\sqrt{K_s K_e}$ instead of C_{66} results in only minor changes in most of the various quantities derived
from the experimental data.

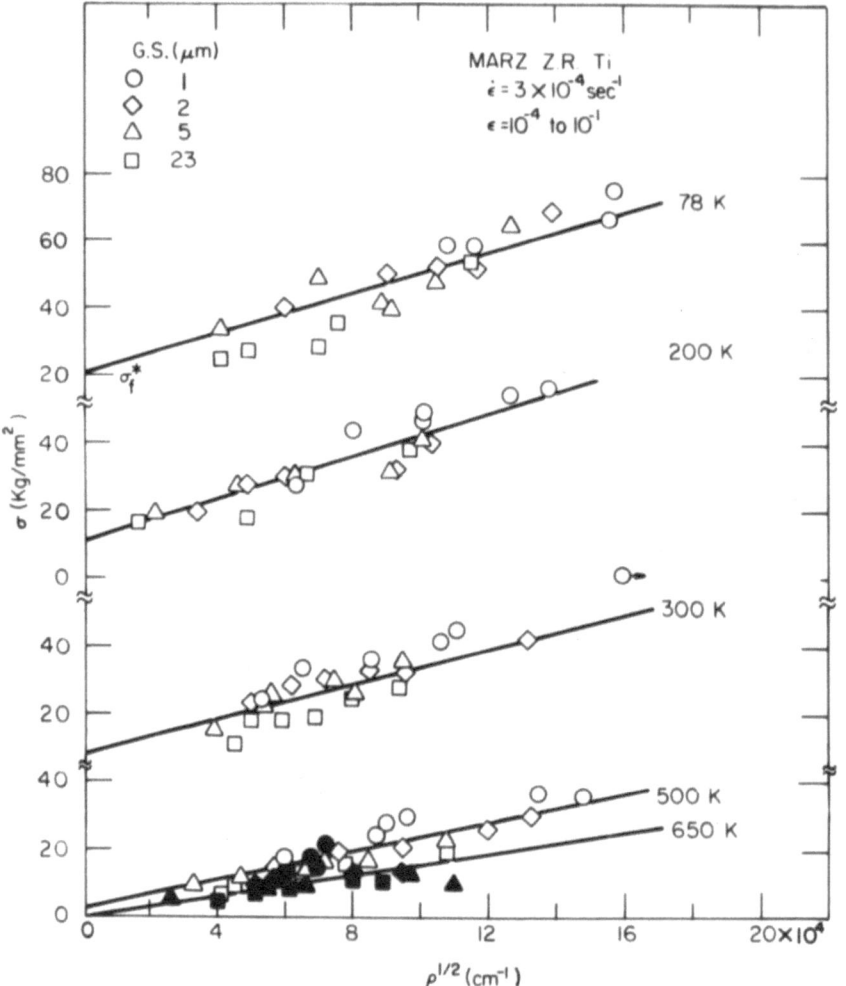

Fig. 2. Flow stress <u>versus</u> the square root of the dislocation density as a function of grain size.

Burgers vector, and ρ the total dislocation density. Hence, one
can rewrite Eqn. (1) as

$$\sigma = \sigma_f^* (T, \dot{\epsilon}, C_i) + 2.0 \ \mu b \rho^{1/2} \ . \tag{2}$$

Related to the temperature-dependence of the flow stress is its
strain rate dependence. The change in stress $\Delta\sigma(\dot{\epsilon})$ resulting from
5:1 changes in strain rate during the straining of the Ti-interstitial

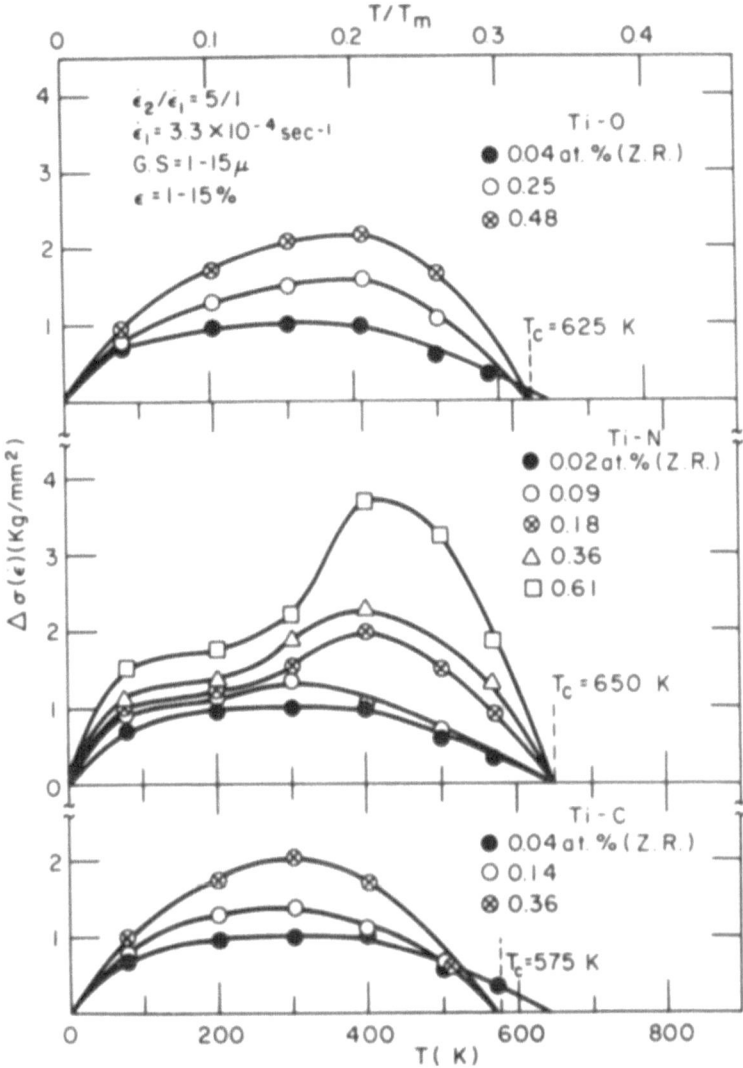

Fig. 3. Effect of temperature on the change-in-stress associated with a 5:1 change in strain rate as a function of interstitial content.

alloys of Fig. 1 is plotted in Fig. 3 as a function of temperature. It was found [7-9] that at a constant temperature $\Delta\sigma(\dot{\varepsilon})$ was essentially independent of strain and grain size, but increased with interstitial content, similar to σ^*. The temperature at which $\Delta\sigma(\dot{\varepsilon})$ extrapolates to zero at high temperatures is in good accord with the value of T_c determined from the flow stress versus temperature curves of Fig. 1, i.e., the lowest temperature where $d\sigma/dT = (\sigma/\mu)\ d\mu/dT$.

In Fig. 4 it is seen that the values of σ^* and $\Delta\sigma(\dot{\varepsilon})$ for the plastic flow of zone-refined polycrystalline wire specimens[†] are in good agreement with the values of τ^* and $\Delta\tau(\dot{\gamma})$ obtained from the

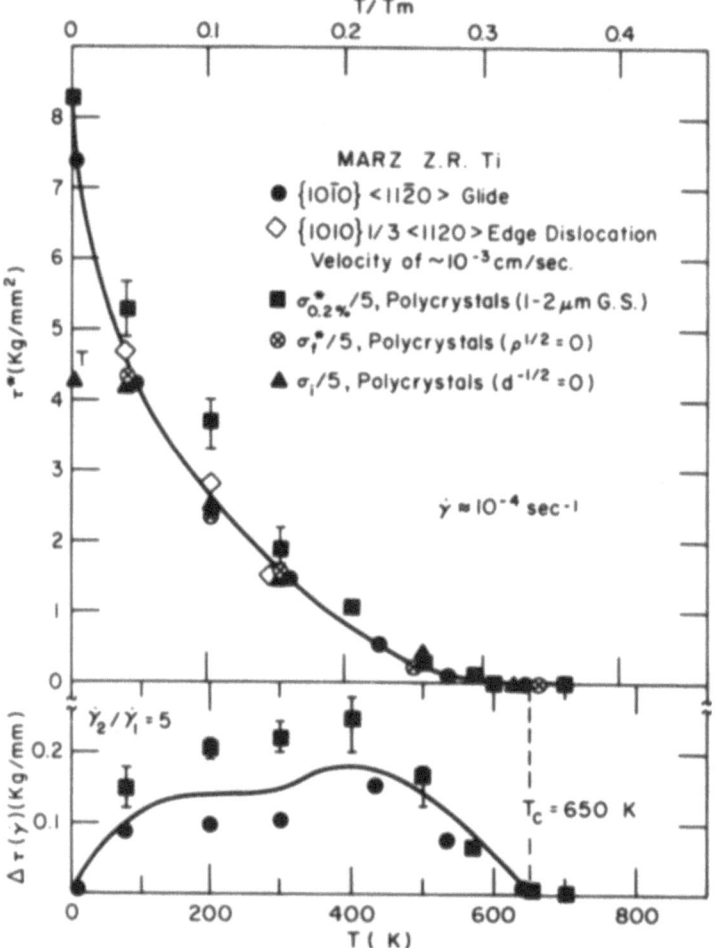

Fig. 4. A comparison of τ^* and $\Delta\tau(\dot{\gamma})$ for prism glide with values obtained from polycrystalline data using a Taylor factor of 5. The symbol T indicates extensive twinning occurred.

[†] The values for σ_f^* were taken from Fig. 2 and those for σ_i^* from Hall-Petch plots of $\sigma_{0.2\%}$ versus the reciprocal of the square root of the grain size [5].

resolved shear stress τ_{RSS} for $\{10\bar{1}0\}$ $<11\bar{2}0>$ slip in single crystals
if one divides the polycrystalline values by 5, i.e., by taking
$\tau^* = 1/5$ σ^*. Moreover, the resolved shear stress required for a
velocity of $\sim 10^{-3}$ cm/sec by $\{10\bar{1}0\}$ $\frac{1}{3}$ $<11\bar{2}0>$ edge dislocations (determined by etch pits) agrees with the critical resolved shear
stress for slip at a strain rate of $\sim 10^{-4}$ sec^{-1} [33,34]. The results
of Fig. 4 thus suggest that the rate-controlling mechanism during the
plastic flow of polycrystals is the same as that during prism slip
in single crystals. Further, the dislocation velocity measurements
suggest that the motion of edge dislocations may be rate-controlling.
However, since the motion of a dislocation occurs by the spreading
of a loop consisting of both edge and screw components, one cannot
decide on the basis of the etch pit results which of the two components is actually rate-controlling. Finally, worthy of mention
regarding Fig. 4 is that the effects of temperature and strain rate
on the resolved shear stress for prism slip presented therein are
in good accord with the results obtained by other investigators on
Ti crystals of similar purity [6,33].

The effect of interstitial solute content on σ^* is given in Fig.
5. Included are the data on wire specimens by Okazaki, et al. [7-9]
in Fig. 1 and those of Finlay and Snyder [1] and Jaffee, et al. [2]
on sheet specimens. Since the total interstitial content of the
starting material used for the sheet specimens [1,2] is nearly the
same as that for the wire specimens [7-9], the same value of σ^* was
used for all materials at 300 K (~ 10 kg/mm^2, derived from Fig. 1).
By so doing, the three sets of data at 300 K are in reasonable agreement for all three interstitial solutes.

To be noted from Fig. 5 is that the curves of σ^* versus C_i are
parabolic in form with an intercept on the stress axis, which increases with decrease in temperature. It is felt that the intercept
reflects mainly the effect of the remaining, residual interstitials,
and hence, that even in zone-refined Ti the interstitial content is
still of sufficient magnitude that the temperature and strain rate
dependence of the flow stress is governed by interstitial solute
obstacles. Of additional interest regarding Fig. 5 is that at 300 K
the strengthening is clearly in the order of C, O, and N, whereas
at 0 K (extrapolated) a difference between C and O is not obvious.

2.3 Solid Solution Strengthening Model

Labusch [35,36] has recently shown that the strengthening which
results from a random array of solutes can be considered to be of
two types: (a) the obstacles to the dislocation motion are mainly
individual solutes (Fig. 6(a)), similar to the model originally
proposed by Fleischer and Friedel [37-39], and (b) the obstacles are
groups of solute atoms (Fig. 6(b)) similar to the model originally
proposed by Mott and Nabarro [40-42]. Further, from both an

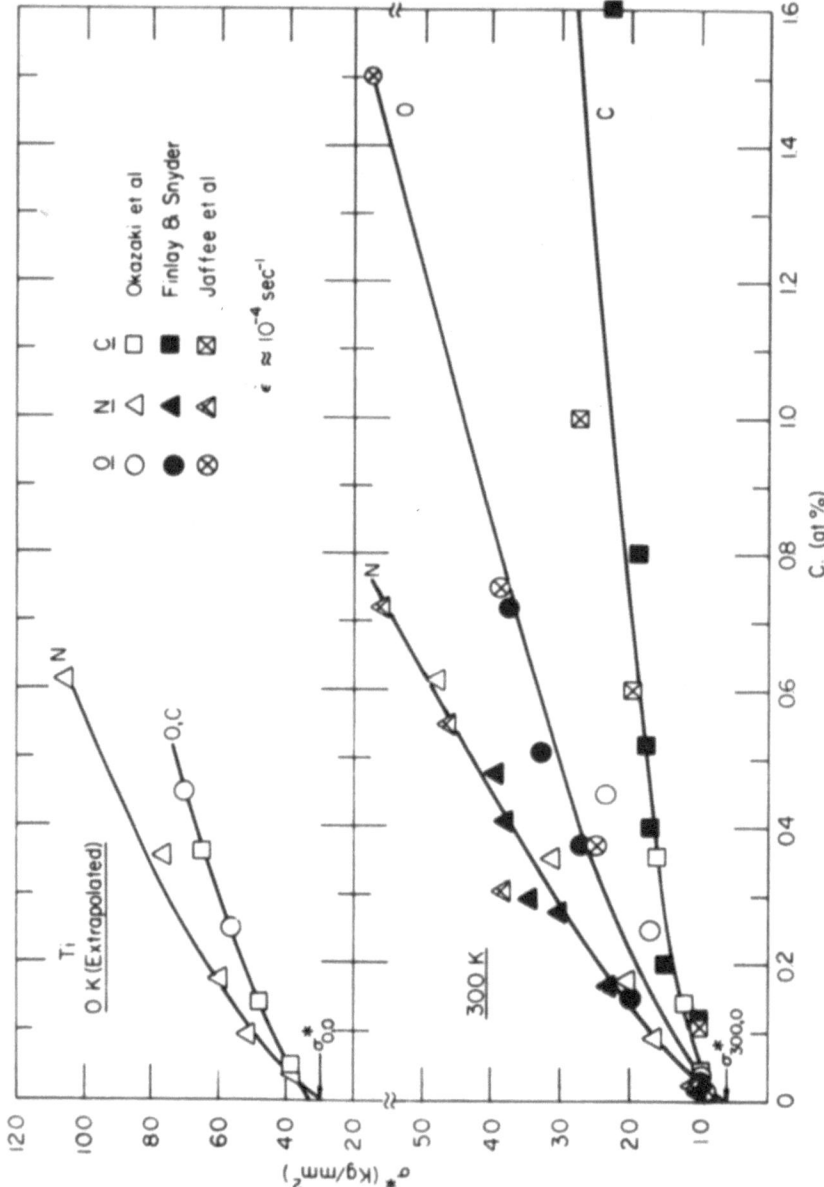

Fig. 5. Effect of interstitial content on σ* at 0 K and 300 K.
Data from Refs. [1,2,7-9].

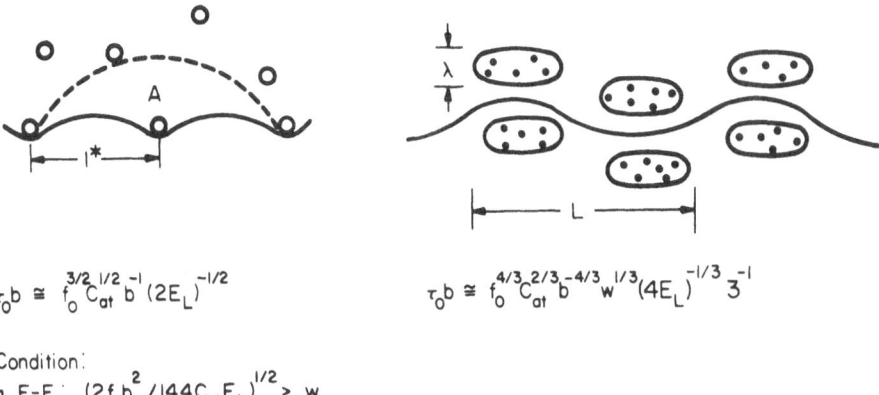

a. Fleisher-Friedel

$$\tau_0 b \cong f_0^{3/2} C_{at}^{1/2} b^{-1} (2E_L)^{-1/2}$$

b. Mott-Nabarro

$$\tau_0 b \cong f_0^{4/3} C_{at}^{2/3} b^{-4/3} w^{1/3} (4E_L)^{-1/3} 3^{-1}$$

Condition:
a. F-F : $(2f_0 b^2 / 144 C_{at} E_L)^{1/2} > w$

b. Labusch : $(f_0 b^2 / 4 C_{at} E_L)^{1/2} > w$

Fig. 6. Solid solution strengthening models, after Labusch [36].

individual process and a statistical analysis, he has shown that the Fleischer-Friedel (F-F) model yields for the flow stress τ_0 at 0 K

$$\tau_0 b \overset{\sim}{=} f_0^{3/2} c_i^{1/2} b^{-1} (2E_L)^{-1/2} \quad , \tag{3}$$

where f_0 is the maximum force of interaction between an individual solute atom and a dislocation, and E_L is the dislocation line tension. For the Mott-Nabarro (M-N) model, Labusch obtains

$$\tau_0 b \overset{\sim}{=} f_0^{4/3} c_i^{2/3} b^{-4/3} (4E_L)^{-1/3} w^{1/3} D \quad , \tag{4}$$

where w is the width of the individual solute obstacle and D is a constant of the order of 1/3. Whether in a given alloy strengthening occurs according to the F-F model or the M-N model will depend on the magnitudes of f_0 and C_i. Labusch concludes on the basis of the amount of bow-out of the dislocation between obstacles that the behavior is described by the F-F model (Eqn. (3)) if

$$(2f_o b^2/144 \ C_i E_L)^{1/2} > w \quad , \tag{5}$$

while on the basis of his statistical analysis, Eqn. (3) describes
the strengthening if

$$(f_o b^2/4C_i E_L)^{1/2} > w \quad . \tag{6}$$

When the parameters on the left side of Eqns. (5) and (6) are less
than w, the solid solution strengthening is given by Eqn. (4).
Thus, both the concentration dependence of the flow stress as given
in Eqns. (3) and (4) and the value of the parameters on the left
side of Eqns. (5) and (6) can be employed to ascertain which of the
two models of Fig. 6 apply to the solid solution strengthening of a
particular alloy.

   Plots of σ^* <u>versus</u> $C_i^{1/2}$ and $C_i^{2/3}$ for the present alloys at 300 K
and 0 K (extrapolated) are given in Figs. 7 and 8. It is seen that
reasonably good straight lines are obtained for both types of plotting, making it difficult to choose between the two. Again, the
relative strengthening due to the three interstitial solutes varies
with temperature. The effect of temperature on the slopes of the
plots of σ^* <u>versus</u> $C_i^{1/2}$ and σ^* <u>versus</u> $C_i^{2/3}$ are presented in Fig. 9.
It is here seen that the strengthening due to N is about 1.5 to 2
times that due to O at all temperatures, while that due to C is not
so clear. It is approximately the same as that due to O in the temperature range of 0 K to 200 K and from 500 to 570 K, but is less
in the intermediate temperature range. Some of this variation may
be due to experimental scatter. In general, it appears that the
strengthening due to C is slightly less than that due to O.

   The values of f_o^*/E_L derived from the data of Figs. 7 and 8 using
Eqns. (3) and (4) above are presented in Table II, along with the
values of the parameters on the left side of Eqns. (5) and (6). It
is seen that in every case the values are higher when derived on
the basis of the M-N model (Eqn. (4)) than the F-F model (Eqn. (3)).
Moreover, it is found that the results meet the requirements of
Eqns. (5) and (6) for the F-F model, although admittedly this is
just on the border line in the case of Eqn. (5). It is therefore
concluded that the strengthening of Ti by interstitial solutes in
the composition range considered (0.1 - 0.6 at.%) may occur by the
interaction of dislocations with individual interstitial solute
obstacles, i.e., according to the F-F model.

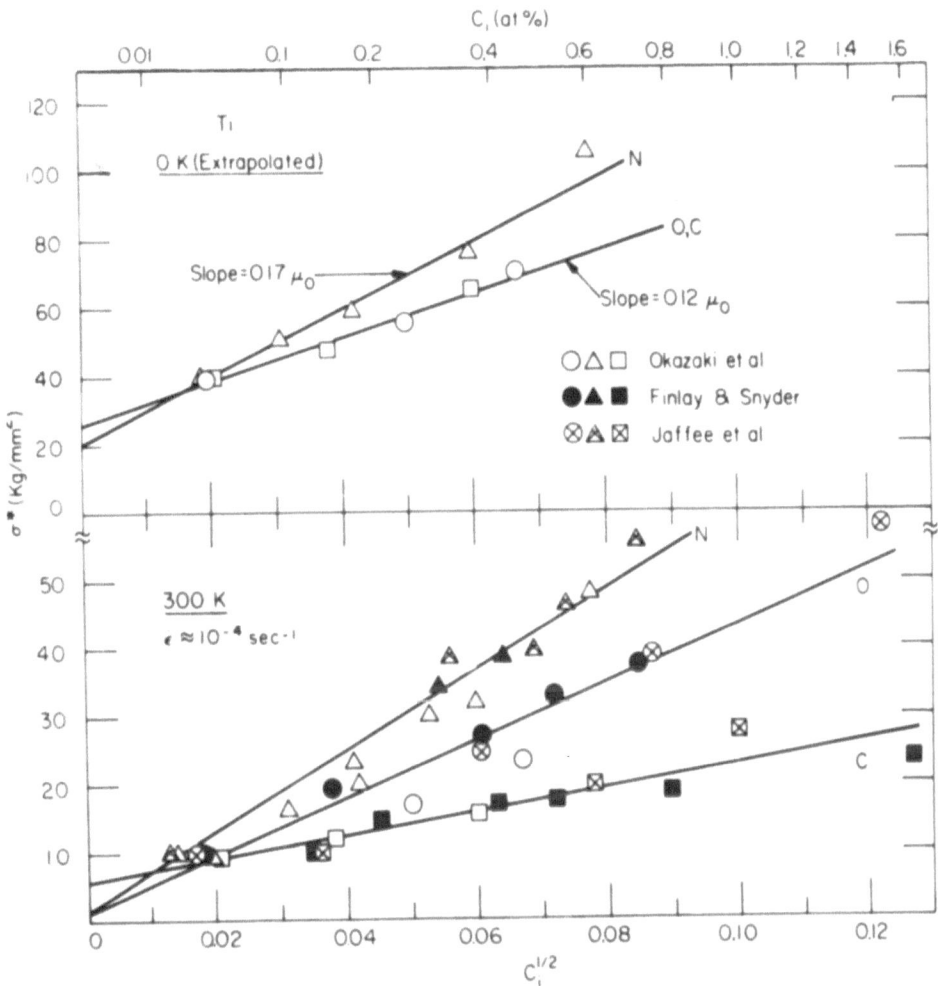

Fig. 7. σ^* versus $C_i^{1/2}$. Data from Refs. [1,2,7-9]

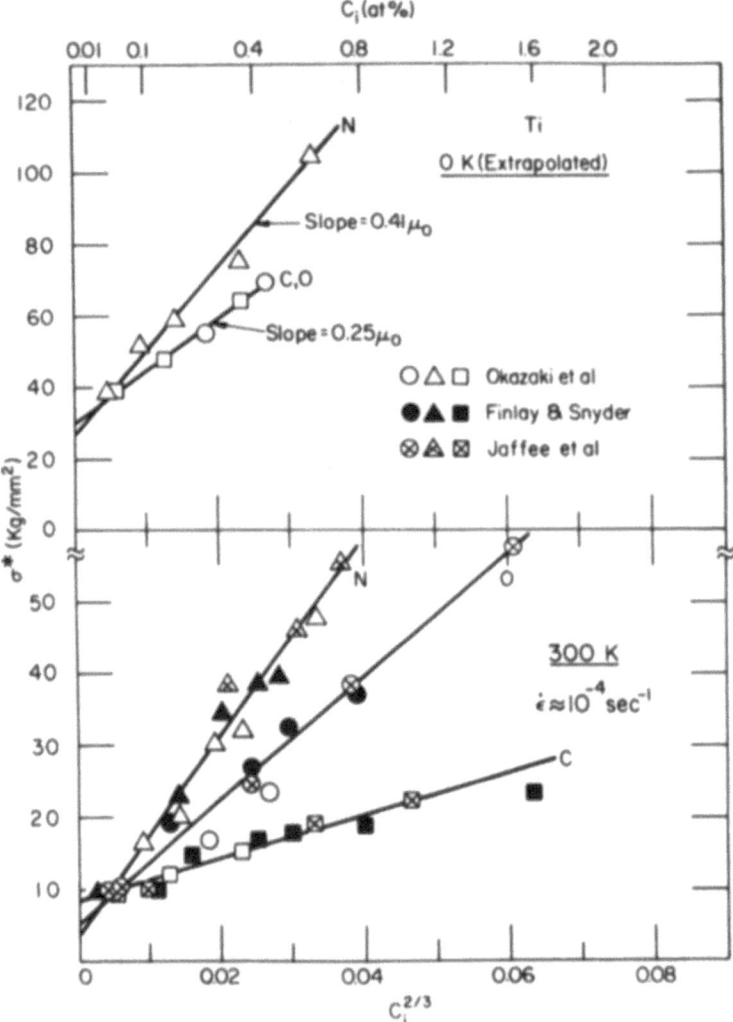

Fig. 8. σ^* <u>versus</u> $C_i^{2/3}$. Data from Refs. [1,2,7-9].

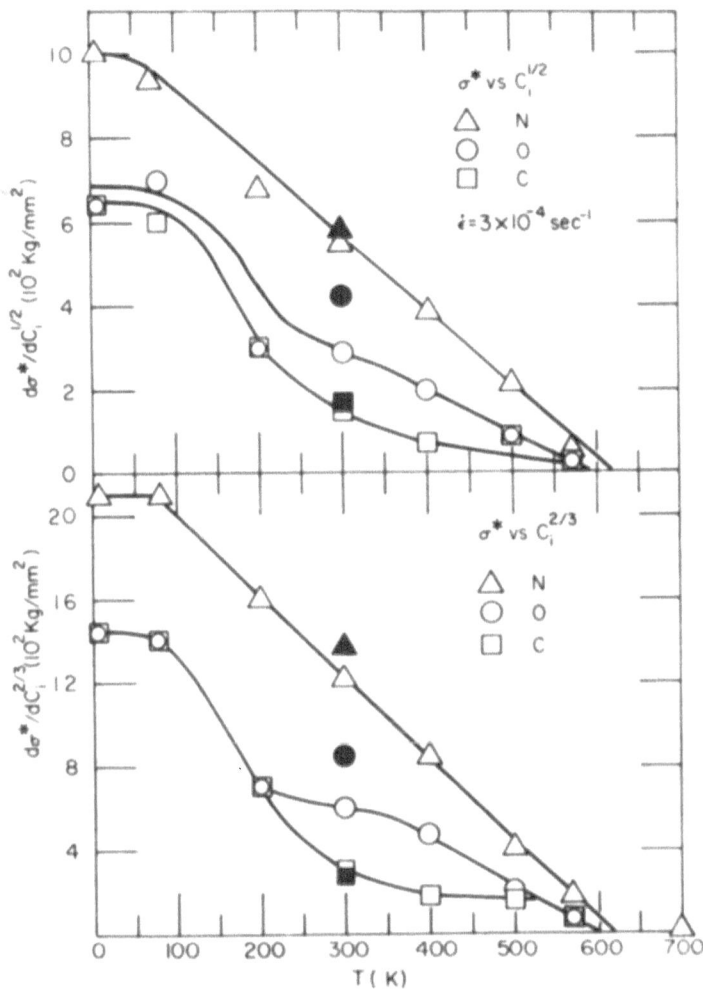

Fig. 9. $d\sigma^*/dC_i^{1/2}$ and $d\sigma^*/dC_i^{2/3}$ <u>versus</u> temperature for Ti-inter-stitial solute alloys. Open symbols are for data from Okasaki, <u>et al</u>. [7-9], filled symbols are for data from Refs. [1,2].

TABLE II.   Parameters Derived from Experimental Data

Alloy System	Composition Range, C_i at.%	f_o^*/E_L Eq. 1 [+]	f_o^*/E_L Eq. 2 [+]	$(2f_o^*/144C_i E_L)^{1/2}$ (Related to Eq. 1) [+]	$(f_o^* b /4C_i E_L)^{1/2}$ (Related to Eq. 2) [+]
Ti-N	0.09 – 0.61	0.16 – 0.19	0.61 – 0.94	0.6b – 1.7b	4.8b – 15.0b
Ti-O	0.25 – 0.45	0.14 – 0.15	0.52 – 0.61	0.7b – 0.9b	5.1b – 7.5b
Ti-C	0.14 – 0.36	0.15 – 0.16	0.58 – 0.74	0.8b – 1.2b	6.0b – 10.7b

[+] Notes:

Eq. 1: $\tau_o^* b = \dfrac{C_i^{1/2} f_o^{*3/2}}{b(2E_L)^{1/2}}$ $E_L = \mu b^2$ (dislocation line tension)

Eq. 2: $\tau_o^* b = \dfrac{C_i^{2/3} f_o^{*4/3} w^{1/3}}{3 b^{4/3} (4E_L)^{1/3}}$ $w = b$ (obstacle width)

$\tau_o^* = \sigma_o^*/5$

3. DEFORMATION KINETICS

One approach to evaluating the effects of temperature and strain rate on the flow stress of crystalline solids is the concept of thermally activated plastic flow. The thermodynamic and physical basis for this approach has recently been reviewed by de Meester, et al. [43]. In this approach the strain rate is expressed in terms of an Arrhenius-type rate equation

$$\dot{\varepsilon} = \dot{\varepsilon}_o \exp \{-\Delta G(\sigma*,T)/kT\} \quad , \tag{7}$$

where $\dot{\varepsilon}_o$ is a constant and ΔG (= $\Delta H*$ - $T\Delta S*$) is the Gibbs free energy of activation, which is a decreasing function of the thermal component of the stress $\sigma*$ (effective stress). $\Delta H*$ and $\Delta S*$ are the enthalpy and entropy, respectively, associated with overcoming of the short-range obstacle to dislocation motion.

Assuming that the F-F model of solid solution strengthening applies, the obstacle spacing $\ell*$ along the dislocation line [44] is given by[+]

$$\ell* = \left(\frac{2M\mu b}{N\sigma*}\right)^{1/3} \quad , \tag{8}$$

where N is the obstacle density on the slip plane, $\sigma*$ (= $\sigma - \sigma_\mu$) is the effective stress and M is the Taylor factor relating the resolved shear stress $\tau*$ to the tensile stress $\sigma*$ (i.e., $\tau* = \sigma*/M$). Further, if a single process is rate-controlling, one can show that [43]

$$\Delta H* = Q - \alpha v\sigma + \frac{3}{2}\alpha v\sigma* \quad , \tag{9}$$

where $Q = - Tv \left(\frac{\partial\sigma}{\partial T}\right)_{\dot{\varepsilon}}$, $v = kT \left(\frac{\partial \ln\dot{\varepsilon}}{\partial\sigma}\right)_T$, and $\alpha = -\frac{T}{\mu}\frac{d\mu}{dT}$.

The variation of $\Delta H*$ (derived using Eqn. (9)) with temperature for $\{10\bar{1}0\}<11\bar{2}0>$ glide in zone-refined single crystals and for the plastic flow of zone-refined polycrystals is presented in Fig. 10.

[+] Kocks [19,45,46] and others [35,47-49] have shown that this relation applies quite well to obstacles of the strength considered here.

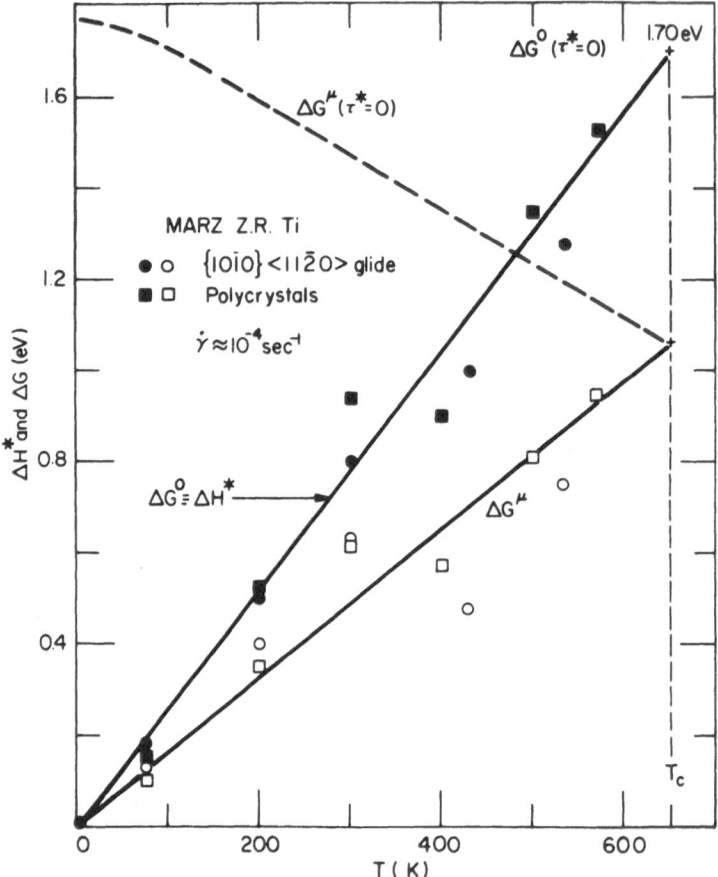

Fig. 10. ΔH* and ΔG as a function of temperature for zone-refined
Ti single crystals and polycrystals.

The data for each material can be considered to yield a straight line
through the origin, indicating that ΔS* is either constant or zero.
Moreover, both the single and polycrystal data lie on a single line,
indicating that the deformation kinetics for the Ti polycrystals are
the same as those for glide on the first-order prism planes. Taking
ΔS* = 0, the Gibbs free energy of activation for the process ΔG⁰
is equal to ΔH*.

On the other hand, Schoeck [50] has proposed that ΔG is proportional to the shear modulus (and hence, also the dislocation-obstacle interaction force). If this is so, then [43,50]

$$\Delta G^{\mu} = (Q - \alpha v\sigma)/(1 + \alpha) \quad . \tag{10}$$

The value of ΔG^{μ} is also plotted in Fig. 10. Within the scatter of the data, ΔG^{μ} can also be considered to be proportional to the temperature (as required by Eqn. (7)), but the fit to a straight line is not as good as it is for ΔG^{O}. However, the value of $\Delta G_{T_c}^{\mu}$ ($\sigma^* = 0$) extrapolated to 0 K by correcting for the change in modulus with temperature is in reasonable accord with the value of $\Delta G_{T_c}^{O}$ ($\sigma^* = 0$), which is independent of temperature.

The effect of temperature on ΔH^* and ΔG^{μ} for the Ti-N alloys is presented in Fig. 11. Again, a proportionality exists between ΔH^* and the temperature; moreover, it is independent of the N content. Also, the data points for ΔG^{μ} do not follow a single straight line as well as do those for $\Delta G^{O} = \Delta H^*$.

Behavior similar to that of Fig. 11 was also found for the Ti-O and the Ti-C alloys. A comparison of the effect of temperature on ΔG^{O} (= ΔH^*) for the C, O, and N alloys is given in Fig. 12. The data points plotted here for each interstitial solute are the averages for the various compositions, since ΔG for a given temperature was independent of interstitial content. To be noted from Fig. 12 is that a single straight line of slope 30.3 k can be drawn through the data points for the three interstitial solutes. The value of ΔG^{O} at $T = T_c$ for the three solutes N, O, and C is 1.73 eV, 1.65 eV, and 1.54 eV, respectively.

Typical variation of the activation volume v with σ^* is illustrated in Fig. 13, which is for the zone-refined polycrystalline material. Here, both v and σ^* have been normalized to 0 K for the two cases considered, i.e., for

$$\Delta G^{O} \quad \left(\text{where } \left.\frac{\partial f_i}{\partial T}\right|_{\tau,x} = 0\right) \quad , \text{ and for}$$

$$\Delta G^{\mu} \quad \left(\text{where } \left.\frac{\partial f_i}{\partial T}\right|_{\tau,x} = -\frac{\alpha f_i}{T}\right) \quad .$$

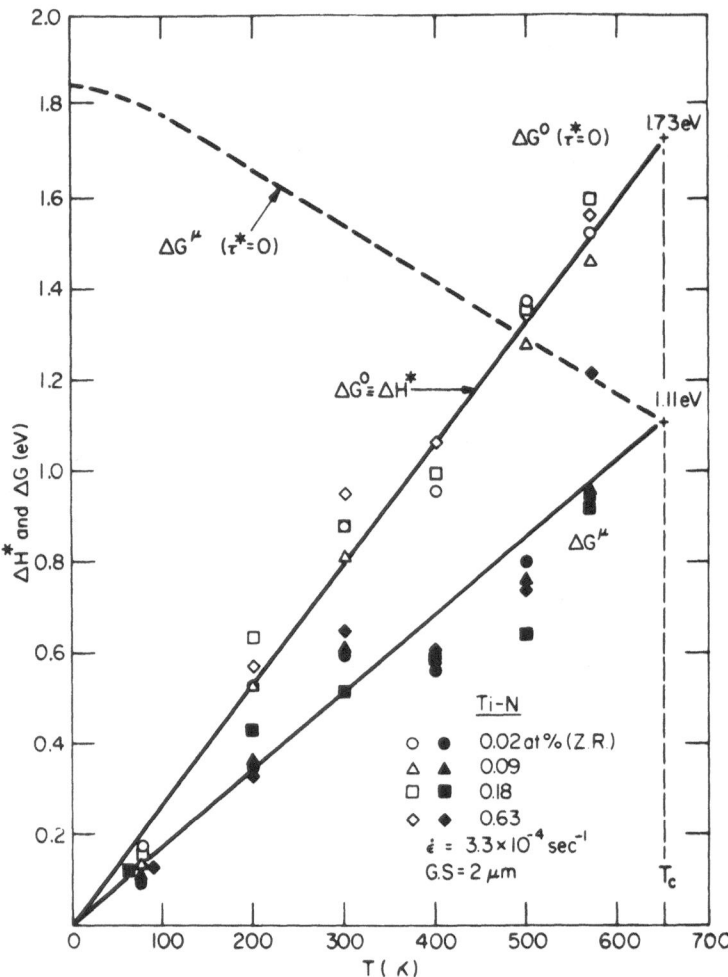

Fig. 11. ΔH^* and ΔG as a function of temperature for Ti-N alloys.

f_i is the interaction force between the dislocation and the obstacle and x is the position of the dislocation.

In the case of $\left.\dfrac{\partial f_i}{\partial T}\right|_{\tau,x} = 0$, the normalization to 0 K consists of

[16,43]

$$\sigma^* \ (0 \ K) = \sigma_T^* \ (\mu_T/\mu_o)^{1/2} \tag{11}$$

and

$$v \ [0 \ K, \ \sigma^* \ (0 \ K)] = v \ (T, \sigma_T^*) \ (\mu_o/\mu_T)^{1/2} \quad . \tag{12}$$

In the case of $\left. \dfrac{\partial f_i}{\partial T} \right|_{\tau, x} = - \dfrac{\alpha f_i}{T}$, the correction to 0 K consists of

$$\sigma^* \ (0 \ K) = \sigma_T^* (\mu_o/\mu_T) \tag{13}$$

and

$$v \ [0 \ K, \ \sigma^* \ (0 \ K)] = v(T, \sigma_T^*) \quad . \tag{14}$$

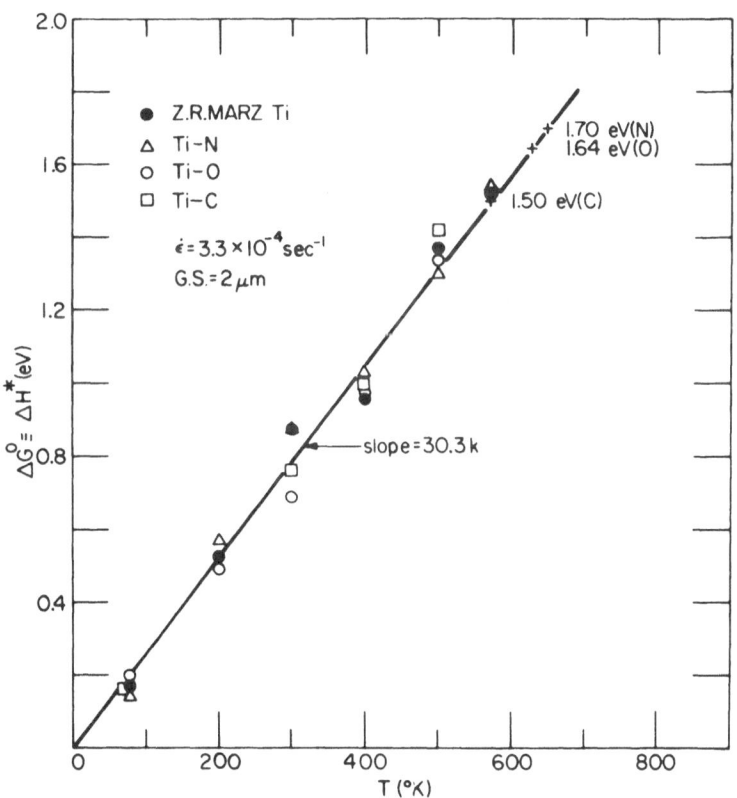

Fig. 12. ΔH^* and ΔG^o as a function of temperature for Ti-C, Ti-N and Ti-O alloys.

Integration of the area under the v <u>versus</u> σ* curve of Fig. 13 should yield ΔG, if a single process is rate controlling over the entire temperature range considered, i.e.,

$$\Delta G = \int_{\sigma^*}^{\sigma_o^*} v d\sigma^* \quad (T = 0 \text{ K}) \quad .$$ (15)

Hence, ΔG derived by means of Eqn. (9) or (10) can be compared with that obtained by Eqn. (15) (all corrected to 0 K) to see whether ΔG⁰ or ΔGᵘ gives the better agreement. This has been done in Fig. 14 for the zone-refined polycrystalline material. It is seen that better agreement occurs for ΔG⁰ (= ΔH*) than for ΔGᵘ, indicating again that the interaction energy is less temperature dependent than is the modulus, and that ΔS* is essentially zero. Furthermore,

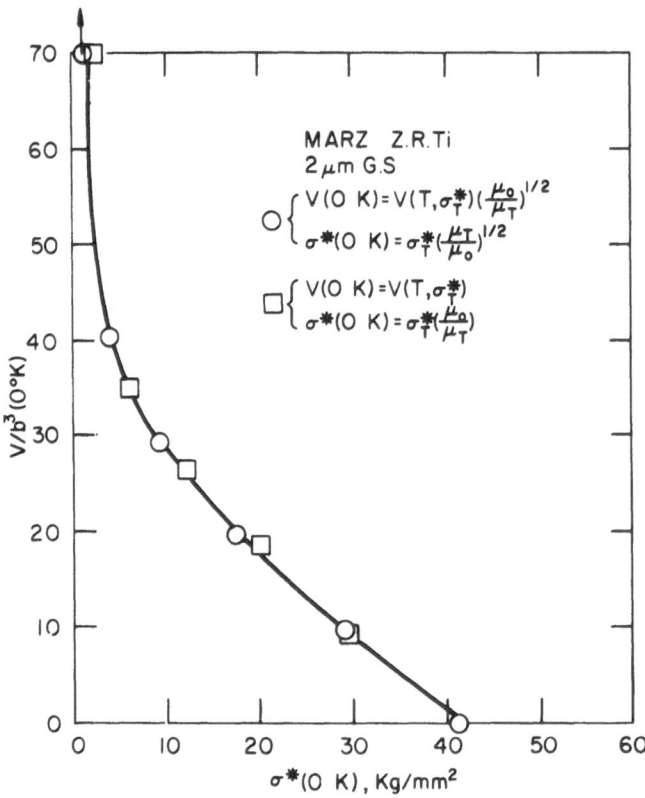

Fig. 13. v/b³ <u>versus</u> σ* corrected to 0 K for zone-refined Ti.

the good agreement between ΔG^o determined from Eqn. (9) at a single temperature and from the integration of the activation volume <u>versus</u> σ^* curve (Eqn. (15)) determined from 0 K to that temperature supports the concept that a single mechanism is controlling over the entire temperature range. Behavior similar to that of Figs. 13 and 14 was found for all of the materials depicted in Fig. 1.

Assuming that the interstitial solutes act individually as obstacles (i.e., the F-F model for solid solution strengthening), we can derive the force-activation distance for their interaction with moving dislocations. The applied force f* pressing the dislocation against the obstacle is given by

$$f^* = \tau^* b \ell^* \quad , \tag{16}$$

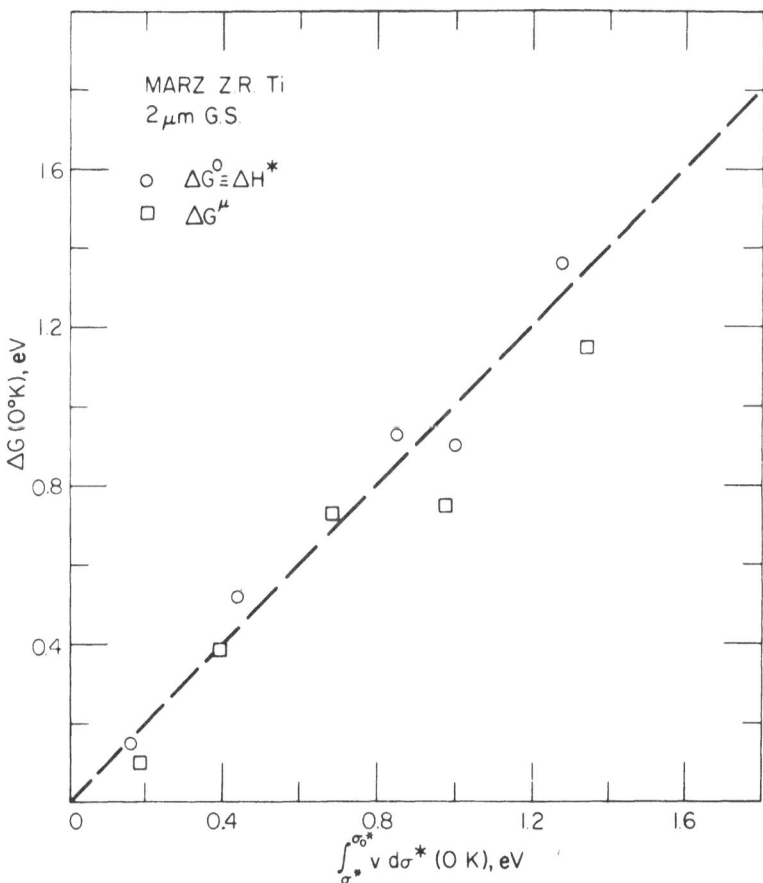

Fig. 14. ΔG <u>versus</u> integration of the area under the v <u>versus</u> σ^* curve (all corrected to 0 K) for zone-refined Ti.

and is exactly balanced by the internal resisting force f_i^* due to
the obstacle. To obtain $\ell*$, we employ Eqn. (8), taking $M = 5$ and

$$N \stackrel{\sim}{=} \frac{C_i}{b^2} , \tag{17}$$

which applies to the density of octahedral sites on one side of the
prism slip plane [51]. Upon inserting these values into Eqn. (16),
one obtains

$$f_i^* = \tau*^{2/3} b^2 (2\mu/C_i)^{1/3} . \tag{18}$$

Further, it can be shown that [43,46]

$$Mv = \frac{2}{3} \ell*bx* , \tag{19}$$

where $x*$ is the activation distance. Inserting the Friedel relation
for $\ell*$ (Eqn. (8)) into Eqn. (19) gives

$$x* = \frac{3Mv}{2b^2} \left(\frac{\tau*C_i}{2\mu}\right)^{1/3} . \tag{20}$$

The force-activation distance curve derived in this manner from
the data on the Ti-N alloys is presented in Fig. 15. The data
points can be considered to fit a single curve independent of interstitial content with a maximum force $f_{i_{max}}^*$ of 0.18 μb^2. Moreover,
the force first rises rapidly at an activation distance of about
1.5 b. Results similar to those of Fig. 15 were also obtained for
the Ti-O and Ti-C alloys, with $f_{i_{max}}^* = 0.17$ μb^2 and 0.16 μb^2, respectively. The results for all three interstitial solutes are given
in Fig. 16, the data points having been eliminated to prevent confusion. It is seen that in general the curve for the Ti-N alloys
lies above that for the Ti-O alloys, which in turn lies above that
for the Ti-C alloys, in accord with the values of the flow stress
and ΔG^O ($\sigma* = 0$).

The results of the deformation kinetics analysis are thus in
accord with the concept that the solid solution strengthening due
to the interstitials C, N, and O in polycrystalline Ti is associated
with the thermally activated overcoming of individual interstitial
solute obstacles by dislocations moving on the first-order prism
planes. Moreover, the deformation kinetics are described reasonably
well by the Arrhenius-type equation with ΔG relatively independent
of temperature and interstitial content.

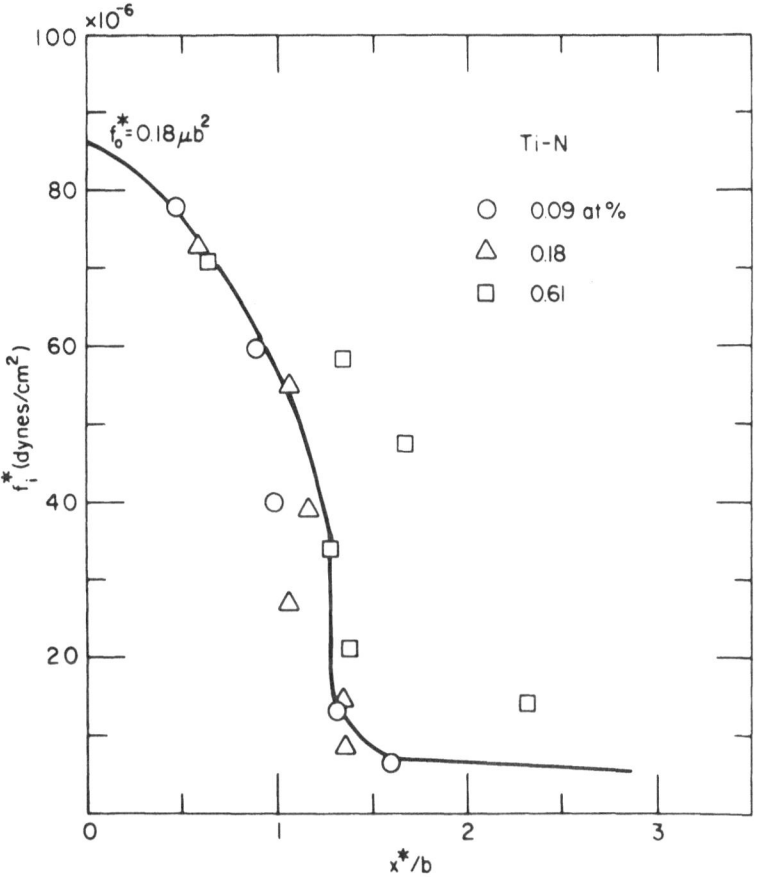

Fig. 15.   Force-activation distance curve for Ti-N alloys.

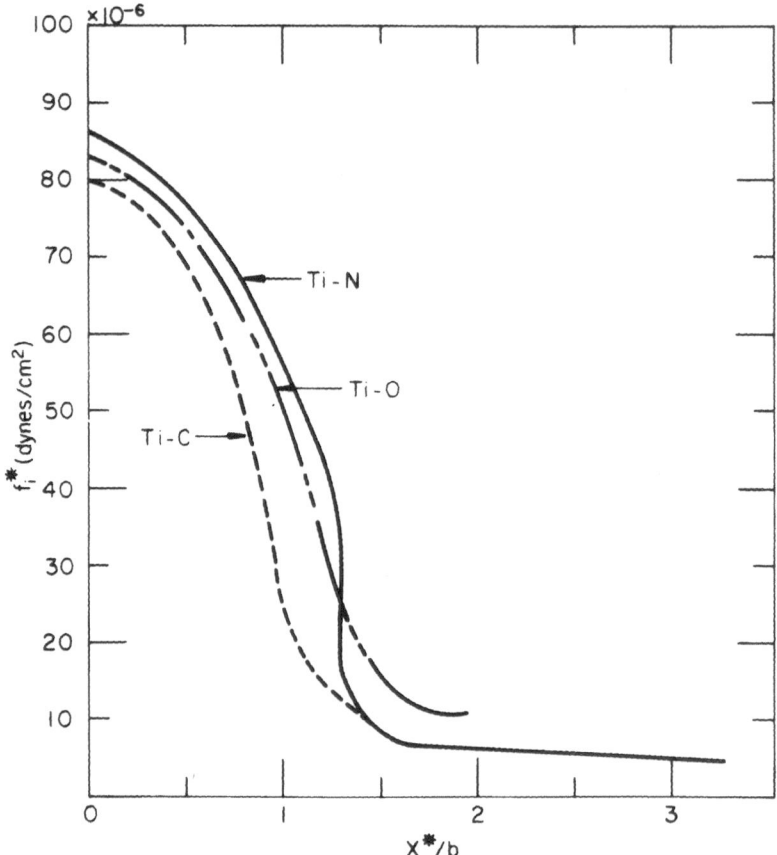

Fig. 16. Force-activation distance curve for Ti-C, Ti-N and Ti-O alloys.

4. NATURE OF THE INTERACTION BETWEEN MOVING
DISLOCATIONS AND INTERSTITIAL SOLUTES

Of interest is the nature of the interaction between moving dislocations and the interstitial solutes which leads to the strengthening discussed in the previous sections. In principle, the interaction energy could be calculated from a knowledge of the electronic rearrangements which occur when a solute atom migrates from some far-removed position in the lattice to one close to the dislocation. However, since the present state of our understanding does not permit such a sophisticated analysis, the approach generally used is to divide artificially the total interaction into a number of pseudo-contributions. These are assumed to be independent and hence can be

added algebraically to obtain the potential of interaction between
the dislocation and a solute atom. The potential so obtained at
constant temperature and pressure is the Helmholtz free energy F*
given by [43]

$$F* = U* - TS*   ,   \tag{21}$$

where U* is the internal energy and S* the entropy. By taking the
derivative of F* with respect to the position x of the dislocation,
we obtain the internal short-range, back-force f_i^* acting on a dislocation segment in contact with an interstitial solute.

The various contributions to the total interaction energy which
have been considered in solid solution strengthening include [37,
38,52-54]:

A. Elastic

    a. Size misfit

    b. Modulus effect

B. Chemical

    a. Phase change

        (1) Associated with a stacking fault

        (2) Associated with the dislocation core

    b. Unlike atom attraction

        (1) Short-range order

        (2) Long-range order

    c. Chemical bonds

        (1) Directed solute-atom/solvent-atom bonding

C. Electrical

    a. Conduction electron density (electric dipole)

Of these various interactions, the following are estimated to
be small for interstitial solutes in Ti [55], and hence can be considered to only make a minor contribution to the large strengthening
which is observed: (a) the phase change associated with a stacking

fault, (b) short-range or long-range order, and (c) an electrical
interaction. This leaves the elastic interactions and the two interactions involving only the dislocation core, i.e., a phase change
at the core and the distortion or breaking of chemical bonds. These
three types of interaction will now be considered for interstitial
solutes in Ti.

4.1 Elastic Interaction

(a) Size Misfit: It is well known that interstitial solutes
in hcp Ti occupy the octahedral interstitial hole positions and
produce a tetragonal distortion of the lattice [56]. A point defect
with tetragonality will interact with both the hydrostatic and
deviatoric components of the stress field. Hence, in general
interstitial solutes will have a large misfit size interaction with
both screw and edge dislocations. However, in the case of $\frac{1}{3}$ <11$\bar{2}$0>
screw dislocations gliding on the first-order prism planes in the hcp
lattice, the interaction energy with defects which produces a tetragonal distortion oriented along the c-axis is zero [57]. Nevertheless,
some size-misfit interaction energy will occur between a screw dislocation and an interstitial solute in Ti due to the volume expansion associated with a screw dislocation [55].

(b) Modulus Effect: Fleischer [37,38] was among the first to
show that an interaction energy will occur between both edge and
screw dislocations and a solute atom due to the local change in
modulus associated with the solute. It is expected that the interaction energy due to this cause will be significant for interstitials in Ti.

(c) Combined Size Misfit and Modulus Effect Interactions: A
detailed analysis of the interaction energy between both $\frac{1}{3}$ <11$\bar{2}$0>
edge and screw dislocations gliding on the first-order prism planes
with interstitial solutes in Ti has been carried out by de Meester
[55]. The results obtained are summarized in Table III. Eshelby
[58] has shown that, since the size misfit and modulus effect
interactions are both elastic, they can be added algebraically.
Hence, the algebraic sum of the two interaction energies is also
given in Table III for both edge and screw dislocations. As mentioned earlier, the interaction energies so calculated are the
Helmholtz free energies F*. Moreover, the derivative of F* with
respect to the position of the dislocation at constant temperature
and pressure is equal to the internal back force f_i^* on the dislocation. The expressions for f_i^* derived in this manner are presented
in Table IV.

By taking the derivative of the force with respect to position and setting this equal to zero, one can obtain the values of the maximum forces and the positions at which they occur. Such a calculation yields that the maximum occurs at $x = y/\sqrt{3}$. At that value of x,

$$\frac{x}{r^4} = \frac{x}{(x^2 + y^2)^2} = \frac{3\sqrt{3}}{16y^3} \quad .$$

TABLE III

Interaction Energies of Interstitials Solutes with Dislocations Gliding on the First-Order Prism Planes in α-Ti.[†]

	Edge dislocation	Screw dislocation
Size	$\pm \dfrac{\mu b}{\pi(1-\nu)} \dfrac{y}{r^2} \varepsilon_e V_{cell}$	$-\dfrac{\mu b}{6\pi^2} \dfrac{K(1+\nu)}{(1-2\nu)} \dfrac{1}{r^2} \varepsilon_s V_{cell}$
Modulus	$\dfrac{\mu b^2}{8\pi^2(1-\nu)^2} \dfrac{1}{r^2} \eta' V_{cell}$	$\dfrac{\mu b^2}{8\pi^2} \dfrac{1}{r^2} \eta' V_{cell}$
Total	$F_e^* = E_e (\eta' \pm \alpha_e \varepsilon_e)$	$F_s^* = E_s (\eta' - \alpha_s \varepsilon_s)$
	$E_e = \dfrac{\mu b^2}{8\pi^2(1-\nu)^2} \dfrac{1}{r^2} V_{cell}$	$E_s = \dfrac{\mu b^2}{8\pi^2} \dfrac{1}{r^2} V_{cell}$
	$\alpha_e = 8\pi(1-\nu) \dfrac{y}{b}$	$\alpha_s = \dfrac{4K(1+\nu)}{3(1-2\nu)}$
	$\varepsilon_e = \varepsilon_b + \nu\varepsilon_c$	$\varepsilon_s = 2\varepsilon_b + \varepsilon_c$

[†] see footnote, Table IV

TABLE IV

Forces of Interaction Between Interstitial Solutes and Dislocations
Gliding on the First-Order Prism Planes in α-Ti.[†]

	Edge dislocation	Screw dislocation
Size	$\pm \dfrac{2\mu b}{\pi(1-\nu)} \; \dfrac{xy}{r^4} \; \varepsilon_e \; V_{cell}$	$-\dfrac{\mu b^2 K(1+\nu)}{3(1-2\nu)\pi^2} \; \dfrac{x}{r^4} \; \varepsilon_s \; V_{cell}$
Modulus	$\dfrac{\mu b^2}{4\pi^2(1-\nu)^2} \; \dfrac{x}{r^4} \; \eta' \; V_{cell}$	$\dfrac{\mu b^2}{4\pi^2} \; \dfrac{x}{r^4} \; \eta' \; V_{cell}$
Total	$f_i^e = f_o^e \, (\eta' \pm \alpha_e \varepsilon_e)$	$f_i^s = f_o^s \, (\eta' - \alpha_s \varepsilon_s)$
	$f^e = \dfrac{b\mu^2}{4\pi^2(1-\nu)^2} \; \dfrac{x}{r^4} \; V_{cell}$	$f^s = \dfrac{\mu b^2}{4\pi^2} \; \dfrac{x}{r^4} \; V_{cell}$
	$\alpha_e = 8\mu(1-\nu) \; \dfrac{y}{b}$	$\alpha_s = \dfrac{4K(1+\nu)}{3(1-2\nu)}$
	$\varepsilon_e = \varepsilon_b + \nu\varepsilon_c$	$\varepsilon_s = 2\varepsilon_b + \varepsilon_c$

Notes: x = distance from the interstitial measured along the slip
plane
y = distance of the interstitial from the slip plane
$r = x^2 + y^2$

$$\varepsilon_b = \left(\frac{\Delta a}{a}\right)_{C_i=0.5} \quad ; \quad \varepsilon_c = \left(\frac{\Delta c}{c}\right)_{C_i=0.5} \quad ;$$

$$\eta' = \frac{\eta}{1 + |\eta/2|} \quad ; \quad \eta = \frac{1}{\mu} \cdot \frac{d\mu}{dC_i} \quad .$$

[†]   negative force of interaction is attractive

The value of V_{cell} in the equations of Table III and IV is $\dfrac{b^2 c_i}{2}$ in accord with Tyson's [57] earlier consideration, since there are two octahedral sites per cell. The assignment of a realistic value to y, the distance the interstitials lie above or below the slip plane, is a more difficult question, because in the hcp lattice the first-order prism planes consist of two corrugated layers of atoms, see Fig. 17. From purely geometrical considerations based on a hard

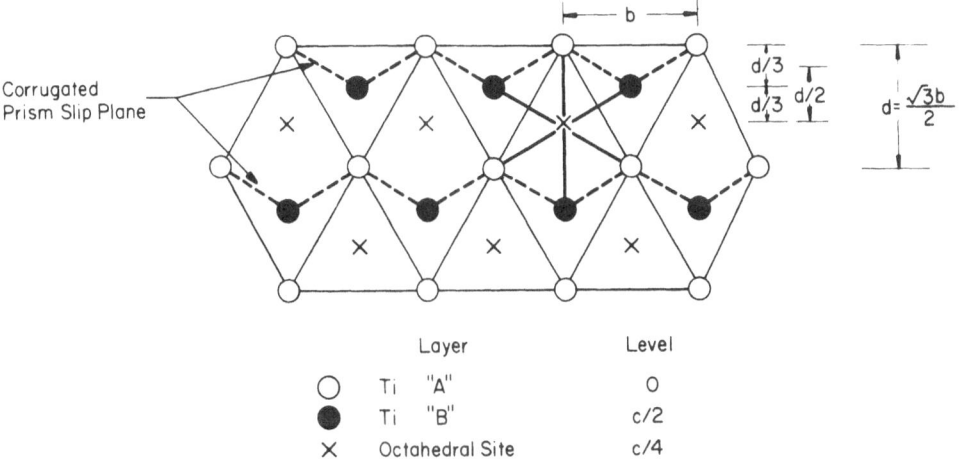

Fig. 17. Position of atoms and distances in the hcp structure, looking down on the basal plane.

sphere model, the average center line of the "thick" plane yields $y = b(\sqrt{3}/2)/2 = 0.43b$. On the other hand, the shortest distance gives $y = b(\sqrt{3}/3)/2 = 0.29b$. In the consideration of Table V below, the value of y which gives the largest interaction energy or force will be used.

The values of the significant parameters in Tables III and IV,

$$\varepsilon_b = \left(\frac{\Delta a}{a}\right)_{c_i = 0.5} \quad ; \quad \varepsilon_c = \left(\frac{\Delta c}{c}\right)_{c_i = 0.5} \quad ;$$

and $\eta' = \dfrac{\eta}{1 + |\eta/2|}$, where $\eta = 1/\mu \dfrac{d\mu}{dc_i}$,

TABLE V. Comparison of the Effect of Interstitial Solute Content on σ_o^* with Size Misfit and Modulus Effect Parameters

Alloy System	$\frac{1}{\mu}\frac{d\sigma_o^*}{dC^{\frac{1}{2}}}$	$\frac{1}{\mu}\frac{d\sigma_o^*}{dC^{2/3}}$	ε_e	ε_s	η'	$|\eta' - \alpha_e \varepsilon_e|$	$|\eta'| + |\alpha_e||\varepsilon_e|$	$|\eta' - \alpha_s \varepsilon_s|$	$|\eta'| + |\alpha_s||\varepsilon_s|$
Ti-N	0.17	0.41	0.040	0.106	-1.275	1.468[†]	1.565[††]	1.463	1.463
Ti-O	0.12	0.25	0.025	0.072	1.325	1.204[†]	1.506[††]	1.197	1.453
Ti-C	0.12	0.25	0.083	0.222	1.395	0.994[†]	1.997[††]	1.000	1.790

$\varepsilon_e = \varepsilon_b + \nu\varepsilon_c$

$\varepsilon_s = 2\varepsilon_b + \varepsilon_c$

$\eta' = \dfrac{\eta}{1 + |\eta/2|}$

$\alpha_s = \dfrac{4K(1+\nu)}{3(1-2\nu)}$

$= 16/9 \ (K=1/3)$

[†] $\alpha_e = 8\pi(1-\nu)\dfrac{y}{b}$

$= 4.837 \ (y=b\sqrt{3}/6)$

[††] $\alpha_e = 7.255 \ (y=b\sqrt{3}/4)$

derived by de Meester [55] from data available in the literature
are presented in Table V. Also presented are the slopes of the plots

of σ^* <u>versus</u> $C_i^{1/2}$ and σ^* <u>versus</u> $C_i^{2/3}$. It is seen that there only

exists a correlation between the concentration dependence of the flow
stress and the combined parameter $|\eta' - \alpha_e \varepsilon_e|$ or the parameter
$|\eta' - \alpha_s \varepsilon_s|$. This correlation is illustrated in Fig. 18.

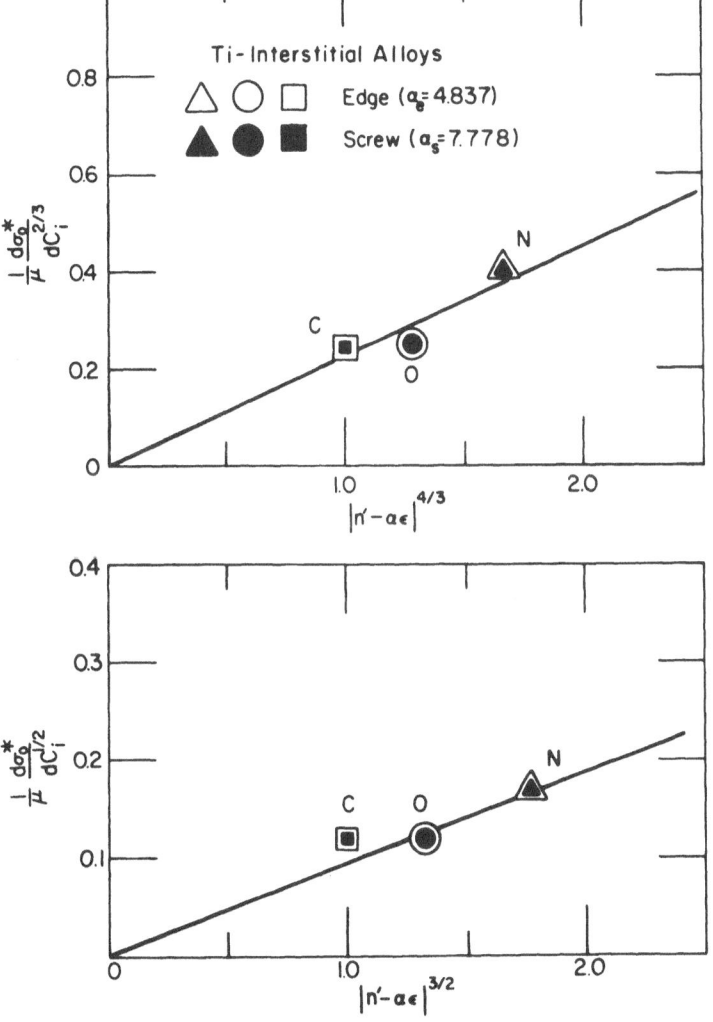

Fig. 18. $1/\mu \ d\sigma_o^*/dC_i^{1/2}$ and $1/\mu \ d\sigma_o^*/dC_i^{2/3}$ versus the interaction
parameter $|\eta' - \alpha\varepsilon|$ for edge and screw dislocations.

The correlation of $\dfrac{1}{\mu}\dfrac{d\sigma^*_o}{dC_i^{1/2}}$ and $\dfrac{1}{\mu}\dfrac{d\sigma^*_o}{dC_i^{2/3}}$ with the parameter

$|\eta' - \alpha_s \varepsilon_s|$ for screw dislocations is reasonable, since the size interaction ε_s with a screw dislocation is negative, whereas the modulus interaction η' may be positive or negative. Hence, the controlling interaction will be given by the algebraic sum of the two. On the other hand, in the case of edge dislocations the value of ε_e may be positive or negative depending whether the solute atom lies above or below the slip plane. Hence, it is expected that the controlling interaction will be the one for which the sum of the two interaction parameters is largest, i.e., for $|\eta'| + |\alpha_e| \, |\varepsilon_e|$. In this regard, Labusch [35] has shown that in the case of edge dislocations, the average interaction force due to the positive and negative interactions on the two sides of the slip plane is given by $f_a^2 + f_b^2$, where $f_a = K_1(\eta' - \alpha_e \varepsilon_e)$ and $f_b = K_1(\eta' + \alpha_e \varepsilon_e)$. This yields $|\eta'^2 + (\alpha_e \varepsilon_e)^2|$ as the significant parameter, which would show a similar variation with interstitial solute as the parameter $|\eta'| + |\alpha_e| \, |\varepsilon_e|$. This all suggests that the correlation of $|\eta' - \alpha_e \varepsilon_e|$ with the concentration of dependence of σ^* may be fortuitous.

A comparison of the value of the interaction energy ΔG and force f_i^* derived from the experimental measurements (maximum values and at $x^* = b$) with those predicted from the combined effect of the size misfit and modulus effect is given in Table VI. It is noted that there exists rather good qualitative agreement between the values calculated and those derived from the experimental data based on both the F-F and M-N models of solid solution strengthening. The agreement is certainly as good as can be rightfully expected, since the values of ΔG and f_i^* at $x^* = b$ and the maximum energy and force are within the core region of the dislocation, where Hooke's law no longer applies. However, as Fleischer [59] has pointed out, the agreement may be better than expected because of some cancellation of errors. For example, the assumption that the stress is constant over the defect will lead to the calculated stress being smaller than would arise by considering the stress variation over the defect. On the other hand, Hooke's law is not expected to apply at distances less than 2.5b from a dislocation and hence, the use of linear elasticity gives a stress that is higher than the true one.

A comparison of the ratio $f_i^*/f_{i\,max}^*$ versus x^*/b derived from the experimental data employing the F-F model with that predicted theoretically on the basis of the equations given in Tables III and IV is

TABLE VI. Comparison of Theoretical Calculations of ΔG and f_i^* Based on the Parameter $|\eta' - \alpha\epsilon|$ with Experimental Data

System	Dislocation	$\Delta G/\mu b^3$				$f_i^*/\mu b^2$			
		$x^*=b$		$x^*=0$		$x^*=b$		$x^*=0$	
		$y=b\sqrt{3}/4$	$y=b\sqrt{3}/6$	$y=b\sqrt{3}/4$	$y=b\sqrt{3}/6$	$y=b\sqrt{3}/4$	$y=b\sqrt{3}/6$	$y=b\sqrt{3}/4$	$y=b\sqrt{3}/6$
Ti-N	Expt'l	0.04[†]		0.19[†]		0.11[†]		0.18[†] 0.16 - 0.19[††] 0.61 - 0.94[†††]	
	screw	0.013	0.014	0.079	0.177	0.021	0.025	0.118	0.399
	edge	0.030	0.031	0.189	0.399	0.053	0.057	0.284	0.898
Ti-O	Expt'l	0.04[†]		0.18[†]		0.10[†]		0.17[†] 0.14 - 0.15[††] 0.52 - 0.61[†††]	
	screw	0.012	0.011	0.065	0.145	0.017	0.021	0.097	0.326
	edge	0.022	0.025	0.138	0.327	0.037	0.047	0.207	0.736
Ti-C	Expt'l	0.08[†]		0.17[†]		0.05[†]		0.16[†] 0.15 - 0.16[††] 0.58 - 0.74[†††]	
	screw	0.009	0.009	0.054	0.121	0.014	0.017	0.081	0.273
	edge	0.015	0.021	0.096	0.270	0.026	0.038	0.144	0.608

Notes:

[†] From plot of f_i^* vs x^*/b (Fig. 16)

[††] $\tau_o^* b = C_i^{\frac{1}{2}} f^{*3/2} / b(2E_L)^{\frac{1}{2}}$

[†††] $\tau_o^* b = C_i^{2/3} f_o^{*4/3} w^{1/3} \Big/ 3b^{4/3} (4E_L)^{1/3}$

$E_L = \mu b^2$

$w = b$

presented in Fig. 19. By plotting the theoretical results in this
manner, one eliminates the differences between edge and screw dislocations and the manner in which the size misfit and modulus effect
parameters are combined. Again, from Fig. 19 it is seen that there
exists a qualitative agreement between the theoretical and experi-

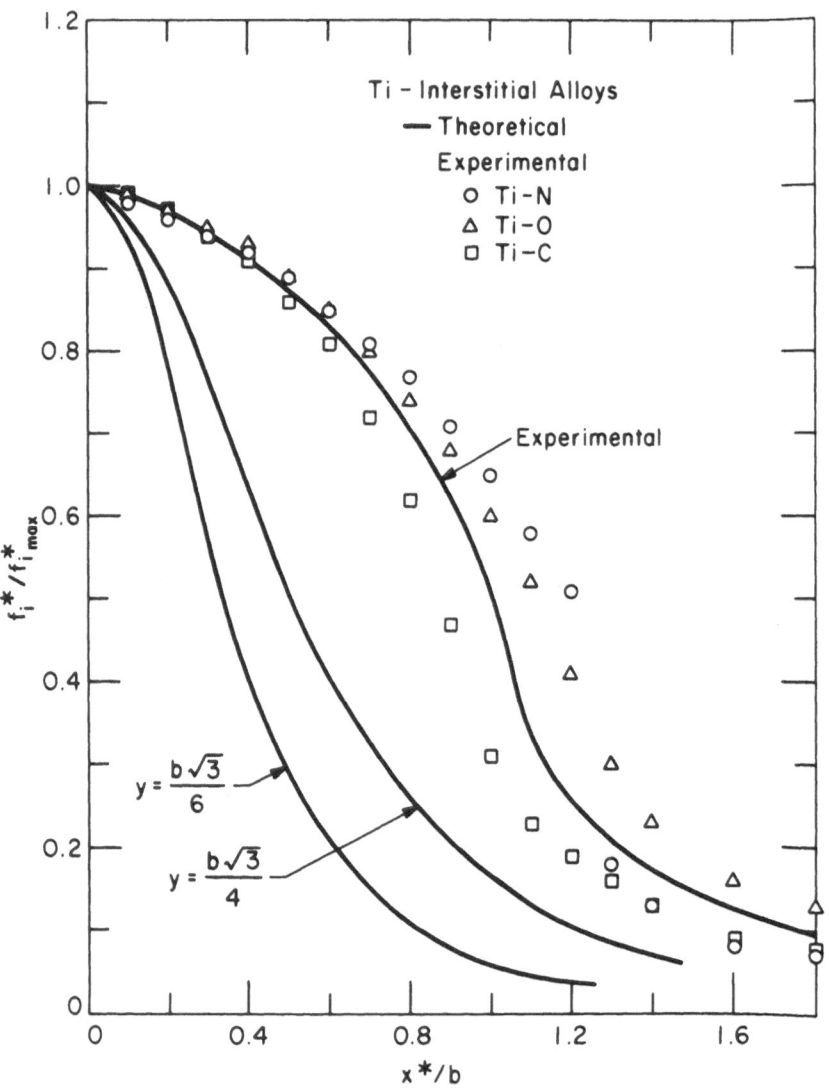

Fig. 19. Force-activation distance curves derived theoretically
on the basis of size misfit and modulus mismatch compared with that
derived from experimental data.

mental curves, the difference between them being within the range of
what one might expect on the basis of the assumptions made. Worthy
of note in this regard is that Kratochvil and Conrad [60] obtained
a reasonably good fit between the experimentally derived f_i^* - x^*
curve for the deformation of Ti-O alloys and that obtained by using
an equation for the interaction energy of the form $F^* = Ab^2y/2(x^2+y^2)$,
which is similar to those employed here. A good fit was obtained by
them upon taking $A = 82 \times 10^{-5}$ dynes/cm^2 and $y = 0.53b$, which values
were derived by comparing the observed variations of σ^* with temperature and v with σ^* with those predicted theoretically employing the
usual thermally activated dislocation motion concept.

Of further interest is a comparison of the observed strengthening due to the presence of two or more of the interstitial solutes
in combination with that predicted theoretically. In this regard,
Kocks [19] has pointed out that for two species of obstacles three
cases are of interest: (a) when $f_{i,1}^* \gg f_{i,2}^*$ and $N_1 \overset{>}{\sim} N_2$, (b)
$f_{i,1}^* \gg f_{i,2}^*$ and $N_1 \ll N_2$, and (c) $f_{i,1}^* \simeq f_{i,2}^*$ and $N_1 \simeq N_2$, where
f_i^* is the force exerted by the obstacle on the dislocation and N is
the area density of the obstacle. In the first case $\tau^* = \tau_1^*$; in the
second, $\tau^* = \tau_1^* + \tau_2^*$; and the third, $\tau^{*2} = \tau_1^{*2} + \tau_2^{*2}$. Computer
experiments by Foreman and Makin [61] have confirmed these relations
and the ranges in between.

Since from the above it appears that the values of f_i^* for C,
N, and O in Ti are not too different, and since the concentration
of these interstitial solutes in Ti are often of similar magnitude,
a reasonable approximation to the combined effect of these interstitials would seem to be

$$\tau^{*^2} = \tau_O^{*^2} + \tau_N^{*^2} + \tau_C^{*^2} \quad . \tag{22}$$

If we assume that the F-F model of solid solution strengthening
applies and we insert for each value of τ^* in Eqn. (22) the value
given by Eqn. (3), we obtain

$$\tau_O^* = [\beta f_{o,O}^{*3} C_O + \beta f_{o,N}^{*3} C_N + \beta f_{o,C}^* C_C]^{1/2} \tag{23}$$

$$= \beta^{1/2} f_{o,O}^{*3/2} (C_O + X_N C_N + X_C C_C)^{1/2} \tag{23a}$$

$$= \beta^{1/2} f_{o,O}^{*3/2} O_{eq}^{1/2} , \tag{23b}$$

where τ_O^* is the effective stress at 0 K, $\beta^{\frac{1}{2}} = 1/[b(2E_L)^{\frac{1}{2}}]$,
$X_N = (f_{o,N}^*/f_{o,O}^*)^3$, $X_C = (f_{o,C}^*/f_{o,O}^*)^3$, and O_{eq} is the oxygen equivalent
of the total interstitial content. Further, if we assume that the

controlling interaction is between screw dislocations and individual interstitial solutes, then

$$X_N = [|\eta' - \alpha_s\varepsilon_s|_N / |\eta' - \alpha_s\varepsilon_s|_O]^3 = 1.81 \text{ , and} \qquad (24)$$

$$X_C = [|\eta' - \alpha_s\varepsilon_s|_C / |\eta' - \alpha_s\varepsilon_s|_O]^3 = 0.59 \text{ .} \qquad (25)$$

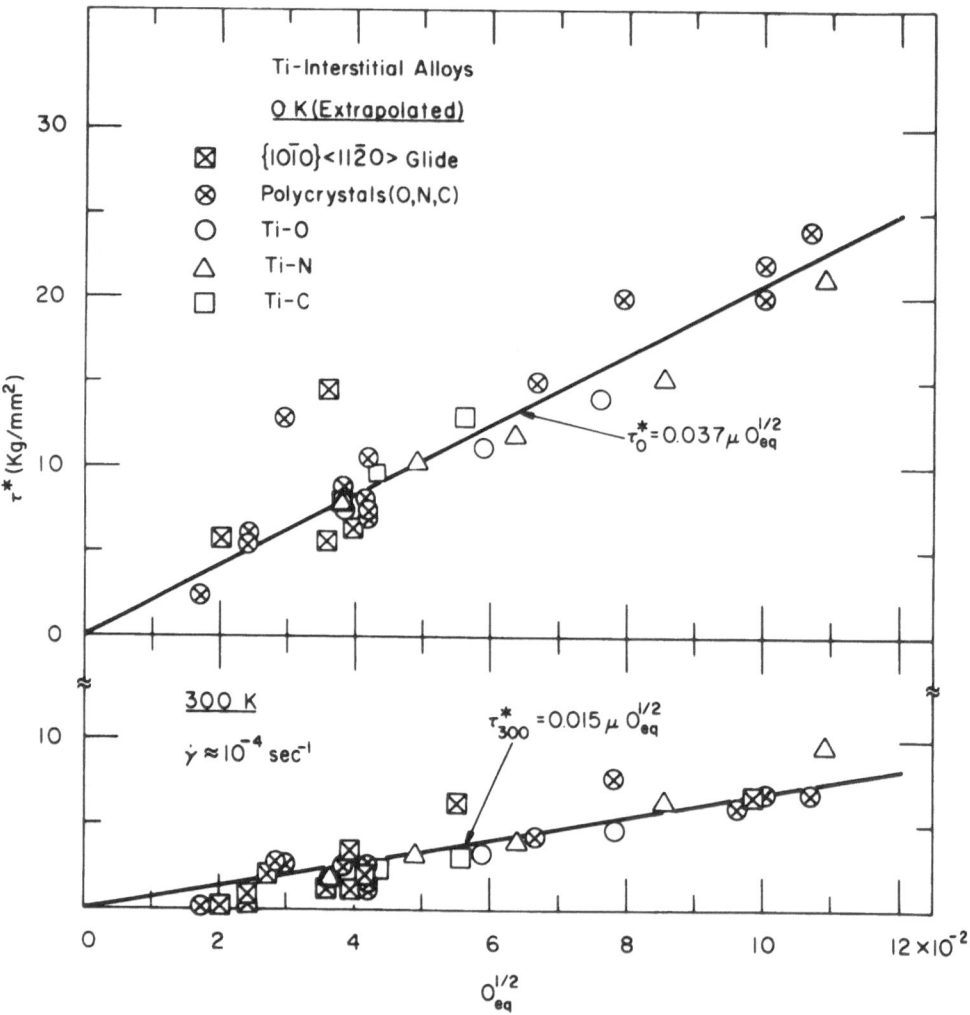

Fig. 20. τ^* <u>versus</u> $O_{eq}^{1/2}$ at 0 K and 300 K based on multiplying factors derived theoretically from the size misfit and the modulus mismatch. Experimental data are from Fig. 1 and sources listed in Ref. [5].

These theoretical values for the multiplying factors needed to obtain an O-equivalent are in good accord with those derived from experimental data by a number of investigators [1,3,55,62-64].

A plot of τ^* at 0 K and at 300 K <u>versus</u> $0_{eq}^{1/2}$ obtained by using the above-listed theoretical multiplying factors on data available in the literature on the critical resolved shear stress for prism slip in single crystals and the yield stress of polycrystals (divided by the Taylor factor of 5) is presented in Fig. 20. It is here seen that the data for the single and polycrystalline specimens of various interstitial contents can be considered to lie on a single straight line through the origin with slope 0.037μ at 0 K and 0.015μ at 300 K. These values of the slopes are essentially the same as those obtained earlier by Okazaki and Conrad [5] using experimentally derived multiplying factors.[†]

Finally of interest is the temperature dependence of F^* and f_i^* (and in turn, ΔG) presented in Tables III and IV. Limited data available in the literature indicate that η increases with increase in temperature [55]. Since μ decreases with increase in temperature, some cancellation will occur due to these opposing effects, yielding a temperature dependence of F^* and f_i^* less than that of μ, in accord with experimental observations. However, before this can be completely accepted, more data on the effect of temperature on η are needed. Also, the effects of temperature on ε_a and ε_c need to be established and taken into account.

All of the above leads to the conclusion that the strengthening due to interstitial solutes during prism glide in Ti single crystals and during the plastic flow of polycrystals is described reasonably well by an interaction between individual solute atom obstacles and screw dislocations moving on the first-order prism planes according to the Fleischer-Friedel model with the interaction energy being given by the combined effect of the size misfit and the modulus change. Moreover, the deformation kinetics associated with this interaction are given by an Arrhenius-type equation where the pre-exponential is relatively independent of stress and temperature and the Gibbs free energy of activation is relatively independent of temperature.

[†] Okazaki and Conrad [5] obtained a value of 0.05μ for the <u>slope</u> at 4.2 K and 0.02μ at 300 K based on $\mu = C_{66}$. Employing $\mu = \sqrt{K_e K_s}$ as was done here, their data yield 0.039μ and 0.015μ, respectively.

4.2 Phase Change Associated with the Dislocation Core

Since it appears from the above discussion that the interaction
between moving dislocations and interstitial solutes occurs mainly
with or very near the dislocation core, interactions other than those
given by the size misfit and modulus change may be important. Regnier
and Dupouy [65] pointed out that the core region of glide dislocations on the prism plane in hcp metals may be considered to have a
bcc structure. The energy of interaction due to this chemical effect
would then consist of the difference in free energy of solution in
the bcc phase as compared to the hcp phase. Since the interstitials
N, C, and O are α-stabilizers, it is expected that their energy would
be higher in the bcc structure of the core and this would then lead
to a short-range repulsive interaction. At present we have insufficient thermodynamic data to evaluate this effect quantitatively. A
rough estimate by Tyson and Conrad [66] for O in Hf gives values
for the interaction energy of the correct order of magnitude.

4.3 Distortion or Breaking of Directed Atomic Bonds

Tyson [57] first proposed that the large energy of interaction
between interstitials solutes and dislocations in Ti might be due
to the breaking of directed atomic bonds between the solute and its
neighbors as the coordination number is changed. Later, Sargent
and Conrad [67] showed that one expects O to be covalently bonded
to Ti in solid solution and that the energy of a single Ti-O bond
should be approximately 1/6 of the heat of formation of TiO (∿8.69
eV). This gives 1.45 eV for the Ti-O bond energy, which is in
reasonable accord with that for plastic flow of Ti-O alloys. Tyson
[54] has more recently calculated similar energies and obtains
values of 2.4 eV, 2.2 eV, and 2.1 eV, respectively, for the Ti-C,
Ti-N, and Ti-O bonds. These again are in reasonable accord with the
measured energies, although the value for Ti-C is out of order with
respect to the others. More work needs to be done to more fully
evaluate this contribution to the interaction energy. That overcoming of interstitial solutes by dislocations gliding on the prism
planes may be considered as the breaking of one of the six nearest
neighbor Ti-interstitial solute atom bonds is illustrated in Fig. 21.

4.4 General Conclusion

From the preceding sections it appears that the major portion
of the interactions between interstitial solutes and dislocations
in Ti occurs at distances very near to or within the dislocation
core. Qualitative, and to a certain degree quantitative, agreement
is obtained between measured values of the interaction energy and
force and those calculated based on a combined effect of size mis-

fit and modulus effect. However, reasonable accord is also obtained
between experimental and calculated values based on thermodynamic
data when the interaction energy is considered to represent the distortion or breaking of covalent-type bonds between the interstitial
solutes and the surrounding Ti atoms. Since the changes in lattice
parameter and modulus which result from the addition of the interstitial solutes to Ti are reflections of the electronic changes
(bonding characteristics) which occur, it could well be that the
calculations based on the size misfit and modulus effect provide a
measure of these electronic energy changes. Thus, the size misfit
and modulus effect parameters may provide a convenient measure of
the more difficult to obtain electronic energy changes. Thermodynamic
data should also be of value in this regard and consideration should
be given to these types of data as well [54,67,69].

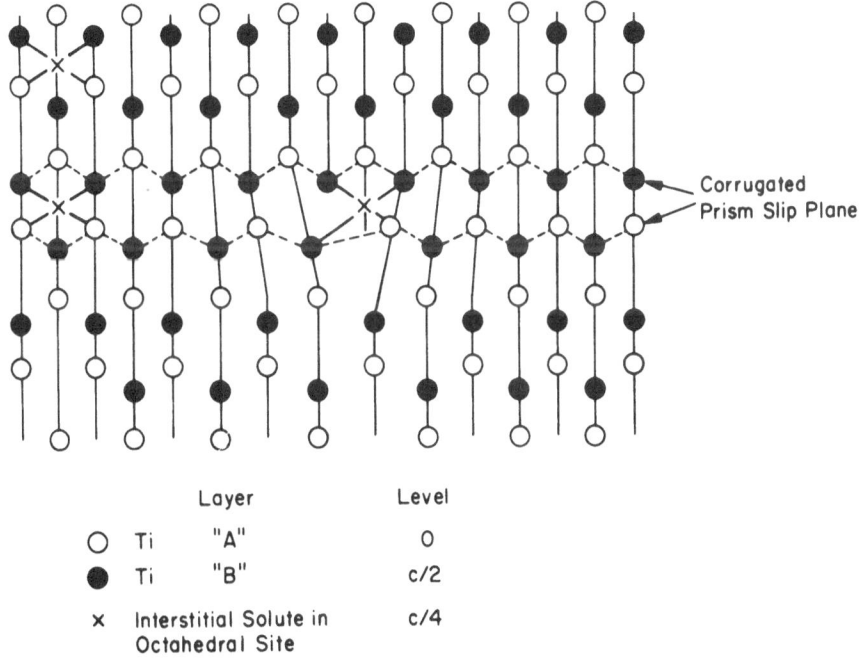

Fig. 21. Edge dislocation on prism plane in hcp structure looking
down on the basal plane. Note that passage of the edge dislocation
breaks one of the six Ti-interstitial solute bonds.

5.   OTHER CONSIDERATIONS

5.1   Preexponential Factor in the Rate Equation

Since the deformation kinetics are described reasonably well by the Arrhenius rate equation (Eqn. 7), it seems desirable to consider the preexponential factor $\dot{\varepsilon}_o$. The slope of the plot of ΔG^o versus T (Figs. 11 and 12) yields $\dot{\varepsilon}_o$ = 4.9 x 10^9 sec^{-1}, whereas that of ΔG^μ versus T (Fig. 11) yields $\dot{\varepsilon}_o$ = 1.5 x 10^5 sec^{-1}. In the F-F model for solid solution strengthening

$$\dot{\varepsilon}_o = (\rho_m/\ell*) \; Ab[(b/\ell*)\nu_D] \; (1/M) \quad , \tag{26}$$

where ρ_m is the mobile dislocation density, A is the area of the glide plane swept out per successful thermal fluctuation, ν_D is the Debye frequency, and M the Taylor factor. Assuming that ρ_m is approximately equal to the total dislocation density at the critical resolved shear stress for prism slip in single crystals or at the 0.2% yield stress in polycrystals, one obtains $\rho_m \cong 10^9$ cm^{-2} from Fig. 2. Further, a reasonable value for $\nu_D \cong 10^{13}$ sec^{-1}, [6]. Inserting these values for ρ_m and ν_D into Eqn. (26) and taking M = 5 and b = 2.95 x 10^{-8} cm gives A = 2.8 x 10^3 $\ell*^2$ for ΔG^o and A = 8.6 x 10^{-2} $\ell*^2$ for ΔG^μ. The large value of A for ΔG^o is in accord with the distance between hard spots ($\sim$10=100 $\ell*$) found by Arsenault and Cadman [68] in their computer experiments on the thermally activated motion of dislocations through a random array of obstacles. On the other hand, the value of this distance calculated from the slope of their plot of $\Delta H*$ versus temperature was only 10^{-2} $\ell*$, in agreement with the value of A obtained here for ΔG^μ. Hence, the computer experiments do not permit a decision regarding ΔG^o versus ΔG^μ.

Taking another approach, one may assume that A = $\ell*^2$ and then calculate ρ_m. The value of ρ_m so obtained is 2.8 x 10^{12} cm^{-2} for ΔG^o and 8.6 x 10^7 cm^{-2} for ΔG^μ. The former value is unreasonably high, whereas this latter is low, but reasonable.

Hence one may conclude that if ΔG is independent of temperature, then for reasonable values of ρ_m the value of A must be greater than $\ell*^2$. If ΔG is dependent on temperature through the modulus, then reasonable values of ρ_m require that A $\overline{<}$ $\ell*^2$.

ACKNOWLEDGMENT

The authors wish to express their appreciation to Mr. L. Rice for assistance in making the dislocation density measurements of Fig. 2 and to Mr. S. Raghuraman and Mr. C. Yin for assistance in carrying out a number of the computations required to obtain the data presented in the text. They also wish to acknowledge support of this research by the Office of Aerospace Research under Air Force Contract F33615-69-C-1027, Mr. A. Adair, technical monitor, and by the Air Force Materials Laboratory under Contract F33615-68-C-1052, Dr. H. Gegel, technical monitor.

REFERENCES

1. FINLAY, W.L. and SNYDER, J.A., Trans. AIME 188 277 (1950).
2. JAFFEE, R.I., OGDEN, H.R. and MAYKUTH, D.J., Trans. AIME 188 1261 (1950).
3. CONRAD, H., Acta Met. 14 1631 (1966).
4. CONRAD, H. and JONES, R., The Science, Technology and Application of Titanium, Pergamon Press, 1970, p 489.
5. OKAZAKI, K. and CONRAD, H., Acta Met. 21 1117 (1973).
6. CONRAD, H., DONER, M. and de MEESTER, B., Titanium Science and Technology, Vol. 2, JAFFEE, R.I. and BURTE, H.M., eds., Plenum Press, 1973, p 969.
7. OKAZAKI, K., MOMACHI, M. and CONRAD, H., Titanium Science and Technology, Vol. 2, JAFFEE, R.I. and BURTE, H.M., eds., Plenum Press, 1973, p 1131.
8. OKAZAKI, K., MORINAKA, K. and CONRAD, H., Trans. JIM 14 470 (1973).
9. OKAZAKI, K., MORINAKA, K. and CONRAD, H., Trans. JIM 15 11 (1974).
10. SANTHANAM, A.T. and REED-HILL, R.E., Met. Trans. 2 2619 (1971).
11. GARDE, A.M., SANTHANAM, A.T. and REED-HILL, R.E., Acta Met. 20 215 (1972).
12. DONER, M. and CONRAD, H., Met. Trans. 4 2809 (1973).
13. OKAZAKI, K. and CONRAD, H., Met. Trans. 3 2411 (1972).
14. OKAZAKI, K. and CONRAD, H., Trans. JIM 13 198 (1972).
15. YIN, C., DONER, M. and CONRAD, H., J. Less-Common Metals 33 229 (1973).
16. de MEESTER, B., DONER, M. and CONRAD, H., Z.f. Metallkde 64 775 (1973).
17. JONES, R.L., Titanium Science and Technology, Vol. 2, JAFFEE, R.I. and BURTE, H.M., eds., Plenum Press, 1973, p 1033.
18. BABEL, H.W. and FREDERICK, S.F., J. Met. (Oct. 1968) p 32.
19. KOCKS, U.F., Physics of Strength and Plasticity, M.I.T. Press, Cambridge, 1969, p 134.
20. FOREMAN, A.J.E., Acta Met. 3 322 (1955).

21. FISHER, E.S. and RENKEN, C.J., Phys. Rev. 135, No. 2A, A482 (1964).
22. CONRAD, H. and OKAZAKI, K., Scripta Met. 4 259 (1970).
23. TYSON, W. and CONRAD, H., unpublished research, Univ. of Kentucky (1972).
24. YIN, C., DONER, M. and CONRAD, H., unpublished research, Univ. of Kentucky (1973).
25. CONRAD, H., Mat. Sci. Eng. 6 265 (1970).
26. CONRAD, H., OKAZAKI, K., GADGIL, V. and JON, M., Electron Microscopy and Structure of Materials, Univ. Calif. Press, Berkeley (1972) p 438.
27. RICE, L. and CONRAD, H., unpublished research, Univ. of Kentucky (1972).
28. WILLIAMS, J.C., SOMMER, A.W. and TUNG, P.P., Met. Trans. 3 2979 (1972).
29. CONRAD, H., Can. J. Phys. 45 581 (1967).
30. OKAZAKI, K. and CONRAD, H., Trans. JIM 13 205 (1972).
31. OKAZAKI, K. and CONRAD, H., Trans. JIM 14 368 (1973).
32. OKAZAKI, K. and CONRAD, H., "Thermal and Athermal Components of the Flow Stress in Dilute Ti-N Alloys at Low Temperatures", unpublished research, Univ. of Kentucky (1973).
33. TANAKA, T. and CONRAD, H., Acta Met. 20 1019 (1972).
34. TANAKA, T. and CONRAD, H., Acta Met. 19 1001 (1971).
35. LABUSCH, R., phys. stat. sol. 41 659 (1970).
36. LABUSCH, R., Acta Met. 20 917 (1965).
37. FLEISCHER, R.L. and HIBBARD, W.R., The Relation Between Structure and Mechanical Properties, N.P.L. Conf. Vol. I, H.M.S.O., 1963, p 262.
38. FLEISCHER, R.L., The Strengthening of Metals, Reinhold Pub. Corp., 1964, p 93.
39. FRIEDEL, J., Electron Microscopy and Strength of Crystals, Proc. 1st. Berkeley Int. Met. Conf., Interscience, 1963, p 634.
40. NABARRO, F.R.N., Proc. Phys. Soc. 58 669 (1946).
41. MOTT, N.F. and NABARRO, F.R.N., Report on Strength of Solids, London Phys. Soc., 1948, p 1.
42. MOTT, N.F., Imperfections in Nearly Perfect Crystals, Wiley, 1950.
43. de MEESTER, B., YIN, C., DONER, M. and CONRAD, H., Rate Processes in Plastic Deformation of Materials, ASM, Cleveland, 1975, p 175.
44. FRIEDEL, J., Dislocations, Addison-Wesley, 1964, p 224.
45. KOCKS, U.F., Can. J. Phys. 45 737 (1967).
46. KOCKS, U.F., Proc. Int. Conf. on the Strength of Metals and Alloys, Suppl. Trans. JIM 9 1 (1968).
47. FOREMAN, A.J.E. and MAKIN, M.J., Phil. Mag. 14 911 (1966).
48. DORN, J.E., GUYOT, P. and STEFANSKI, T., Physics of Strength and Plasticity, M.I.T. Press, Cambridge, 1969, p 133.
49. ARGON, A.S., Phil. Mag. 25 1053 (1972).
50. SCHOECK, G., phys. stat. sol. 8 499 (1965).

51. TYSON, W.R. and CRAIG, G.B., Can. Met. Quart. 7 119 (1969).
52. FLINN, P.A., Strengthening Mechanisms in Solids, ASM, Metals Park, Ohio, 1962, p 17.
53. HAASEN, P., Proc. Int. Conf. on the Strength of Metals and Alloys, Trans. JIM 9 (1968).
54. TYSON, W.R., "Solution Hardening by Interstitials in Close-Packed Metals", paper in this volume.
55. de MEESTER, B., Ph.D. Thesis, Univ. of Kentucky (1972).
56. BISOGNI, E., MAH, G. and WERT, C., J. Less-Common Metals 7 1977 (1964).
57. TYSON, W.R., Can. Met. Quart. 6 301 (1968).
58. ESHELBY, J.D., Solid State Physics, Vol. 3, Academic Press, 1957, p 79.
59. FLEISCHER, R.L., Acta Met. 10 835 (1962).
60. KRATOCHVIL, J. and CONRAD, H., Scripta Met. 4 815 (1970).
61. FOREMAN, A.J.E. and MAKIN, M.J., Can. J. Phys. 45 511 (1967).
62. WOOD, R.A., Titanium Metallurgy Course, New York Univ., College of Engineering, Lecture 4 (Sept. 13-15, 1965).
63. JAFFEE, R.I., Prog. Met. Phys. 7 65 (1968).
64. OKAZAKI, K. and CONRAD, H., Trans. JIM 14 364 (1973).
65. REGNIER, P. and DUPOUY, J.M., phys. stat. sol. 28 K55 (1968).
66. TYSON, W.R. and CONRAD, H., Met. Trans. 4 2605 (1973).
67. SARGENT, G. and CONRAD, H., Scripta Met. 6 1099 (1972).
68. ARSENAULT, R.J. and CADMAN, T., "Thermally Activated Motion Through a Random Array of Obstacles", John E. Dorn Memorial Symposium on Rate Processes in Plastic Deformation, ASM, in print.
69. COLLINGS, E.W. and GEGEL, H.L., "Physical Principles of Solid Solution Strengthening in Alloys", paper in this volume.

SOLUTION HARDENING BY INTERSTITIALS

IN CLOSE-PACKED METALS

W. R. Tyson[†]

Department of Physics, Trent University

Peterborough, Ontario, Canada

ABSTRACT

Solution hardening of hcp Zr, Ti, and Hf by interstitial O and N, and of fcc Ni and Th by interstitial C is examined in the light of existing theories. It is concluded that hardening in the hcp alloys is due to chemical effects, while elastic interactions are significant in the fcc systems.

1. INTRODUCTION

Foreign atoms dissolved in a metal invariably cause changes in its mechanical properties, called "solution hardening" if the alloy atoms remain in solution and no precipitation occurs. The simplest type of solution hardening to treat is that in which solute atoms are distributed at random in the lattice and do not diffuse appreciably at test temperatures, so that moving dislocations interact with stationary barriers in the slip plane. It is this "dispersed barrier hardening" (DBH) phenomenon that will be of interest in this paper. We shall not be concerned with dislocation locking or yield-point phenomena, as these are not significant in the systems we shall discuss.

The fundamental problem of theories of DBH is the calculation of the strain rate $\dot{\gamma}$ as a function of effective shear stress τ^*,

---

[†] Now at Physical Metallurgy Research Laboratories, Department of Energy, Mines and Resources, Ottawa, Canada.

temperature T, and alloy concentration c$^+$. Considerable progress
on this problem has been made by the application of <u>reaction rate
theory</u> to plastic deformation which leads to the well known formula
for the strain rate

$$\dot{\gamma} = \dot{\gamma}_o \, e^{-\Delta G/kT} \quad ,$$
(1)

where $\dot{\gamma}_o$ is the "frequency factor" and ΔG the free energy of activation for the overcoming of barriers. A lucid discussion of the predictions of rate theory for low-temperature deformation has been
given by Dorn [2], and thorough treatments of the thermodynamics of
<u>thermally activated flow</u> are available (de Meester, <u>et al.</u> [3] and
Surek, <u>et al.</u> [4]).

In this paper, we will first review the principal features of
reaction rate theory as it applies to DBH, and discuss the variation
of τ^* with T and c expected under different conditions. The theory
will then be applied to a particularly interesting series of alloys,
namely interstitial solid solutions in close-packed metals. These
alloys are ideal examples of DBH, since interstitials cause appreciable strengthening at low temperatures and the total interaction
energy G_o between dislocations and interstitials is large. This is
in contrast to the case for bcc alloys where the intrinsic lattice
resistance complicates DBH effects in both substitutional and interstitial alloys, and for substitutional alloys in general where G_o
is usually small. Finally, possibilities for the physical origin
of the dislocation-solute interaction (DSI) will be considered.

2. SOLUTION HARDENING THEORIES

A dislocation-solute interaction can be characterized by the
variation of interaction force F with distance x between dislocation
and solute. The F – x profile will generally have a form resembling
that of Fig. 1, which is appropriate for a symmetrical repulsive
interaction; the range of the interaction depends on its physical
origin and on the width of the dislocation. Under an effective
stress τ^*, a dislocation is pressed against an obstacle in its path
and will be thermally activated over it if the temperature is sufficiently high. The details of the processes by which lengths ℓ of

---

+ We shall assume that the flow stress τ may be separated into
$\tau = \tau^* + \tau_\mu$ in the normal way, where the long-range athermal stress
τ_μ varies with the shear modulus μ; this separation seems justified
for the systems we shall be considering (see Conrad [1] for experimental tests in the Ti-O system).

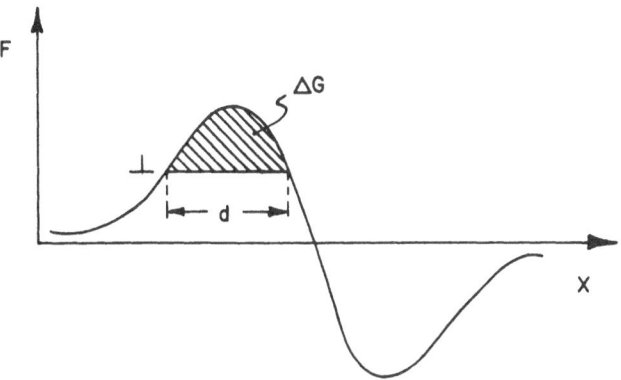

Fig. 1. Schematic variation of force F between dislocation and obstacle with distance x along the slip plane for a symmetrical repulsive barrier. ΔG and d are the activation energy and distance, respectively.

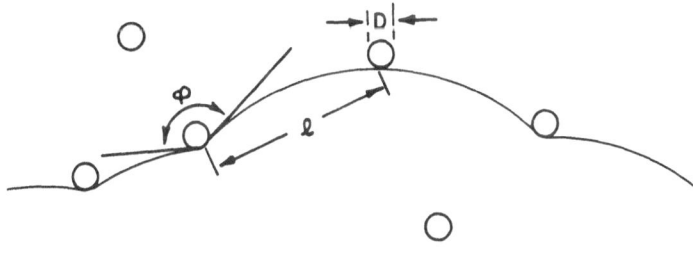

$$F = 2T \cos (\varphi/2)$$
$$\approx \tau^* \ell b$$

Fig. 2. Bowing of a dislocation under an effective stress τ^* against short-range obstacles of width D.

dislocation move in an activated event are of two major kinds:
barriers may be overcome individually, or groups of barriers may be
involved in the activated process. The former case would be expected
for very dilute solutions, while the latter should be the case for
barriers with long-range stress fields, or for concentrated solutions,
where the range of the DSI is comparable to the interparticle spacing. We will discuss theories relating to these two possibilities
in turn.

2.1. Single Barrier Theories

If the region on the slip plane in which the stress field of
the dispersed barriers is significant is only a small fraction of
the total area, then a dislocation can bow out freely between the
obstacles along its length (Fig. 2). The force F is determined by
the angle ϕ at the obstacle:

$$F = 2T \cos (\phi/2) \quad , \tag{2}$$

where T is the line tension of the dislocation. T depends on the
orientation of the dislocation line and on its curvature as discussed
by Brown and Ham [5]; we shall assume the approximate relation
$T = \mu b^2/2$. Under the force F, the dislocation will press against
the barrier as shown in Fig. 1, and will overcome the barrier by
thermally activated motion through the "activation distance" d if
the energy ΔG can be supplied by thermal fluctuations.

In the special case of a regular square array of obstacles, ϕ
will be the same at all obstacles and the activation length ℓ (for
a dislocation moving parallel to a side of the square array) will
be given by

$$\ell = \lambda = 1/\sqrt{n} \quad , \tag{3}$$

where λ is the "square array spacing", and n is the concentration
of obstacles per unit area in the slip plane.† Taking the atomic
concentration $c \sim nb^2$, we have $\lambda \sim b/\sqrt{c}$. (The mean spacing between
particles in three dimensions for a cubic array, however, is
$\Lambda \sim b/c^{1/3}$). Under an effective stress τ^*, the dislocation can bend

---

† For interstitials in octahedral sites in close-packed lattices,
it is easily shown that for hcp lattices $\lambda^2 = a'c'/2c$ for $\{10\bar{1}0\}$
planes and $\lambda^2 = (\sqrt{3}/2) (a')^2/c$ for (0002) planes where a', c' are
lattice parameters and c is the atomic concentration.

For fcc lattices $\lambda^2 = (\sqrt{3}/2) b^2/c$ where $b = a_o \sqrt{2}/2$ is the Burgers
vector (a_o the lattice parameter).

to a radius of curvature $r = T/(\tau^*b)$ and it follows from simple geometry that $F = 2T \cos \phi/2 = \tau^*\lambda b$. The effective stress at 0 K is
therefore

$$\tau_0^* = F_m/(\lambda b) = \mu\alpha\sqrt{c} \quad, \tag{4}$$

where $F_m = \alpha\mu b^2$; for single solute atoms α normally is in the range
0.01 - 0.2 (Fleischer, [6]). At higher temperatures, τ^* decreases
due to thermal fluctuation. In the simplest case, that of a square
force-distance profile, we have $\Delta G = G_0 - \tau^*b\lambda D$ where $G_0 = F_m D$ and
from Eqn. (1)

$$\frac{\tau^*}{\mu} \cdot \frac{\lambda}{b} = \frac{F_m}{\mu b^2} - \frac{1}{(D/b)} \cdot \frac{kT}{\mu b^3} \ln \frac{\dot{\gamma}_0}{\dot{\gamma}} \quad. \tag{5}$$

The frequency factor is given approximately by $\dot{\gamma}_0 = \rho b^2 \nu_D$ where ρ
is the mobile dislocation density and ν_D is the Debye frequency.
This relation is plotted in Fig. 3, which shows the effect of increasing $\dot{\gamma}$ by a factor of 100 and of increasing F_m at constant D/b
by a factor of two. Note that increasing D/b at constant F_m has
the effect of increasing the critical temperature T_c at which
$\tau^* = 0$ by the same factor.

Also shown in Fig. 3 is a dotted curve which applies if the
length ℓ between obstacles is given by the "Friedel relation". As
pointed out by Friedel [7], if the dispersed barriers form a random
array rather than a regular one, the number of obstacles touched by
a given length of dislocation will increase as the dislocation bows
out under increasing effective stress τ^*. A simple derivation†
gives

$$\frac{\ell}{b} = (\frac{\lambda}{b})^{2/3} (\frac{\tau^*}{\mu})^{-1/3} \quad. \tag{6}$$

For bowing between points touching an initially straight dislocation,
we then have $F = \tau^*\ell b$ with ℓ/b given by Eqn. (6). An elegant statistical treatment by Kocks [8] and computer experiments by Foreman
and Makin [9] have confirmed the applicability of Eqn. (6) for bar-
riers having breakthrough angles $\phi \gtrsim 100°$, i.e., $\alpha \lesssim 0.64$; since
$\alpha \leq 0.2$ for single solute atoms, Friedel's relation should apply in
DBH. Since these treatments take into account the fact that ϕ varies
from obstacle to obstacle along the dislocation, it is apparent that
for $\alpha \leq 0.64$, which corresponds to "weak barriers" and quasi-straight

---

† Note that in an earlier paper by Tyson [10], 2μ was written in
Eqn. (6) instead of μ. This is equivalent to taking the line tension as μb^2 rather than $\mu b^2/2$; the latter estimate is more widely
used.

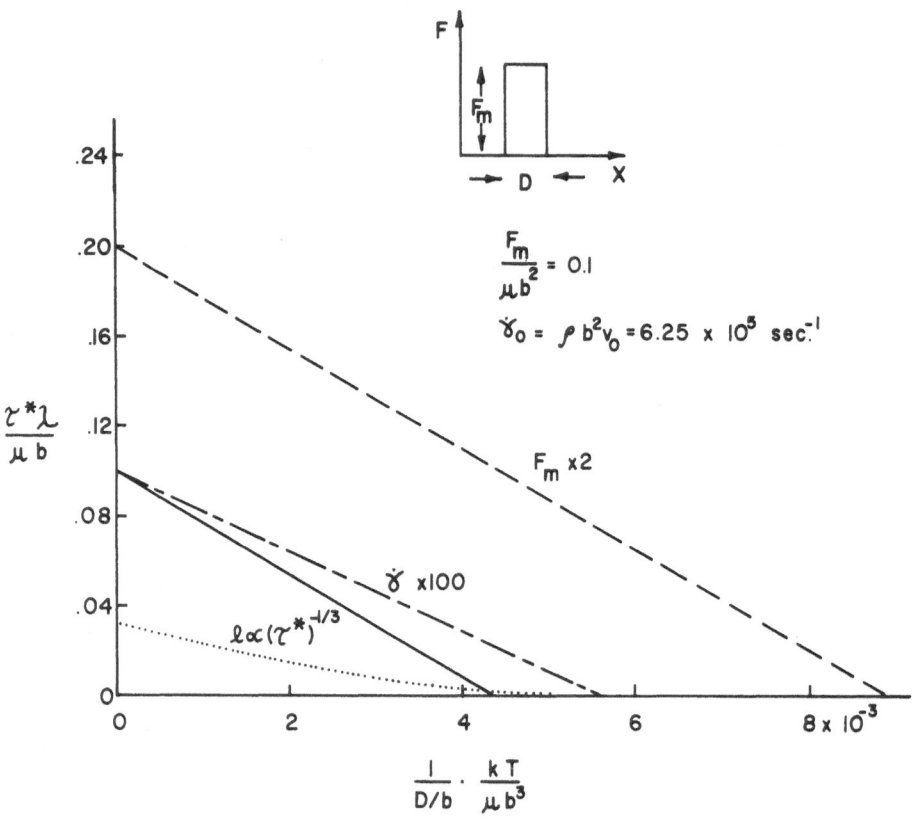

Fig. 3. Variation of effective stress with temperature for a rectangular force-distance profile, showing the effect of increasing
$\dot{\gamma}$ and F_m and allowing for stress dependence of ℓ.

dislocations, the relation $F = \tau^* \ell b$ is a good approximation. Using
the Friedel relation Eqn. (6), the effective stress at 0 K is given
by

$$\tau^*_o = \mu \alpha^{3/2} \sqrt{c} \quad . \tag{7}$$

Since $\alpha \leq 0.64$ for this treatment to hold, comparing Eqns. (4) and
(7), we see that τ^*_o is reduced from the square-array value; this
is consistent with the observation that $\ell > \lambda$ and, since $F = \tau^* \ell b$,
τ^* is therefore smaller at a given F. This effect is evident in

Fig. 3, as is the curvature in the effective stress – temperature curve due to variation of ℓ with τ^*. Also, as pointed out by Dorn [2], the frequency factor $\dot\gamma_0$ is a function of τ^* in this case, and so there is no critical temperature T_c at which $\tau^* = 0$; however, $\dot\gamma_0$ is only a weak function of τ^*, and, since it enters logarithmically in the equation for τ^*, this effect is not of great importance.

Curvature in the effective stress – temperature relation is generally expected, not only because of the variation of ℓ with τ^* but also because of deviations of the barrier profile from a rectangular shape (Ono [11]). As an example, Fig. 4 compares the behavior for a rectangular (R) barrier and a triangular (T) one, assuming for simplicity that ℓ is constant. The curve for the triangular barrier is quite representative of the majority of barrier profiles investigated by Ono.

The square-array spacing Eqn. (3) and the Friedel spacing Eqn. (6) apply to the case of short-range repulsive barriers. For barriers with D (Fig. 3) larger than $\sim$b, or for attractive barriers, other spacings could apply. The mean free length of straight dislocation between barriers of range D is given by $\ell = 1/(2nD)$, i.e.,

$$\ell = b^2/(2cD) \quad . \tag{8}$$

Alternatively, in the case of attractive barriers, Friedel [12] has shown that a dislocation would assume a zigzag shape with the length between barriers given by

$$\frac{\ell}{b} = (\frac{\mu b^3}{U} \cdot \frac{1}{c^2})^{1/3} \quad , \tag{9}$$

where U is the DSI energy. For Eqns. (8) and (9), the flow stress at 0 K is proportional to c and $c^{2/3}$, respectively. However, for these spacings to apply, they must be smaller than the Friedel spacing Eqn. (6). For Eqn. (8), this leads to the condition

$$c > \alpha/(2D/b)^2 \quad ,$$

where $\alpha = F_m/\mu b^2$. With $\alpha = 0.2$, this requires $c > 0.05$ for $2D/b = 2$ and $c > 0.013$ for $2D/b = 4$. Eqn. (8) could therefore only apply for relatively concentrated alloys, in which case the "internal stress theories" of the next section should be more appropriate. In the case of Eqn. (9), the condition is

$$(\frac{1}{c})^{1/3} (\frac{\tau^*}{\mu})^{-1/3} > (\frac{1}{\gamma c^2})^{1/3} \quad ,$$

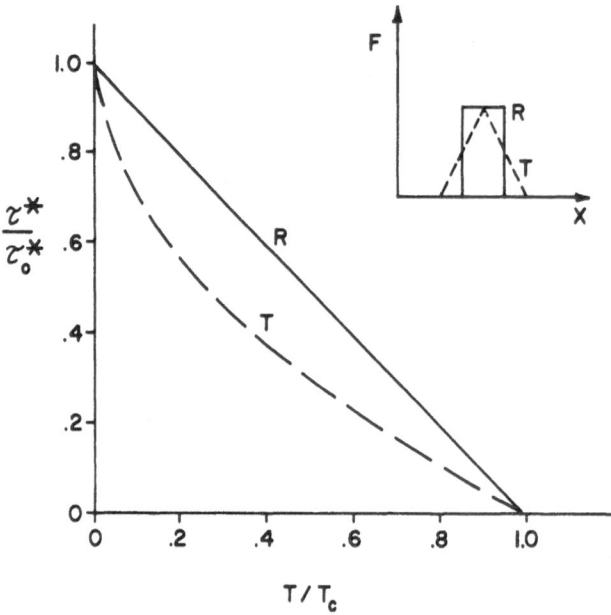

Fig. 4. Variation of normalized effective stress τ^*/τ_o^* with normalized temperature T/T_c for a rectangular (R) barrier profile and a triangular (T) one, assuming ℓ = constant.

where

$$\frac{\tau^*}{\mu} = \frac{F_m}{\mu b^2} \Bigg/ (\frac{\ell}{b}) = \alpha(\gamma c^2)^{1/3} \quad ,$$

and

$$\gamma = U/(\mu b^3) \quad .$$

This leads to $c > \alpha^3/\gamma^2$, i.e., $c \gtrsim \alpha$, since $U \sim F_m b$ and so $\alpha \sim \gamma$. Hence, Eqn. (9) could apply only for very concentrated alloys or very weak barriers ($\alpha \lesssim 0.01$). A similar conclusion was reached by Brown and Ham [5].

2.2 Internal Stress Theories

In alloys containing barriers with a long-range stress field,
i.e., in which the range of the barrier force is comparable to ℓ,
the segments of dislocation between barriers will be subject to an
internal stress field from the dispersed barriers as well as to the
effective stress. If the barrier stress fields are spherically symmetric, it is evident that their wavelength will be of the order of
the volumetric interparticle spacing $\Lambda = b/c^{\frac{1}{3}}$. This wavelength can
be quite small even in fairly dilute alloys, as shown in Table I.
The possibility therefore exists for thermal activation over distances of this order.

Table I.   Interparticle Spacings

c	Planar $\ell/b = c^{-1/2}$	Volumetric $\Lambda/b = c^{-1/3}$
.0001 (.01 at.%)	100.0	21.5
.001 (.1 at.%)	31.6	10.0
.01 (1 at.%)	10.0	4.6
.1 (10 at.%)	3.2	2.2

Mott and Nabarro considered the effects of such relatively
short wavelength[†] internal stresses in a series of papers, lucidly
summarized by Cottrell [13]. The internal stresses, assumed to
be due to the size misfit of solute atoms, vary as $1/r^3$ where
r is the distance from the solute atom. The local curvature of a
segment of length Λ is determined by the volume average of this internal stress field $\tau_i \sim \mu \varepsilon c \ \ell n(1/c)$; the size misfit parameter is
given by $\varepsilon = 1/a \ da/dc$, where a is the lattice parameter. The
theory considers the correlated motion of a length $n^2 \Lambda$ of dislocation between equilibrium positions separated by a distance Λ. The

---

[†] i.e., wavelength $\sim \Lambda$; internal stresses with wavelengths $\gg \Lambda$ (due,
for example, to dislocation pile-ups) cannot be overcome thermally
and so contribute to τ_μ rather than τ^*.

average force to do this is calculated statistically, from the random sum of forces from barriers on both sides of the dislocation, to be

$$\tau_o^* \simeq 2.5 \ \mu\varepsilon^{4/3}c \ .$$ (10)

The activation length is constant for a given concentration in this theory, and the activation energy is $G_o \sim \varepsilon^{2/3}\mu b^3$. Since typically $\varepsilon \sim 0.01$ to 0.2 and $\mu b^3 \sim 5$ eV, we have $G_o \sim 0.2$ to 1.7 eV, which is a range accessible to thermal activation. The variation of effective stress with temperature will, as in the case of interaction with single solute atoms, be determined by the force-distance profile as the length $\ell = n^2\Lambda$ moves between positions of equilibrium.

More recent attempts at the calculation of τ_o^* for particles with long-range strain fields ("coherency strains") have been discussed by Brown and Ham [5], who emphasize the averaging problems involved. However, the essential difference between the "single barrier" and "internal stress" cases is the fact that in the latter instance obstacles near the dislocation line either repel or attract it randomly, so that a statistical problem involving interactions of alternating signs is involved in deducing the flow stress rather than one of simply averaging. This feature is incorporated in the Mott-Nabarro theory, but not in the later approaches reviewed by Brown and Ham.

Labusch [14,15] (see also Nabarro [16]) has reconsidered the calculation of the flow stress in a solution hardened alloy using statistical averaging and Green's function methods. He finds $\tau_o^* \propto c^{1/2}$ or $c^{2/3}$, depending on whether the value of $\beta = \alpha/2cw^2$ is larger or smaller than a critical value of approximate magnitude $\beta \sim 25$. For values of β smaller than $\beta \sim 0.5$, the initial calculations predicted $\tau_o^* = 0$ and special considerations were introduced later [15] to retrieve a finite τ_o^* in this range. In the present state of the theory, the concentration dependence of τ_o^* for $\beta \lesssim 1$ is difficult to deduce with certainty. Direct calculation of the flow stress in computer simulation studies by Ono [17] indicates that the exponent n in $\tau_o^* \propto c^n$ could vary over a range of values with $0.3 \leq n \leq 0.7$ for $0.01 \leq \alpha \leq 0.05$, although these results should be considered tentative due to the assumptions made to simplify the computations.

2.3 Evaluation of Thermodynamic Parameters

The activation energy ΔG (Fig. 1) may be evaluated experimentally from measurements of the dependence of τ^* on T and $\dot{\gamma}$, although assumptions are necessary concerning the temperature dependence of the total activation energy G_o and the stress dependence of ℓ. Also, the strain-rate sensitivity $(\partial\tau/\partial\ln\dot\gamma)_T$ may be used to evaluate the activation area ℓd although, again, assumptions are required about the stress dependence of ℓ. Formulae for ΔG and ℓd have been given by de Meester et al. [3] for $G_o = G_o^o = $ constant, or $G_o = G_o^\mu \propto \mu$, and $\ell = $ constant or $\ell \propto (\tau*/\mu)^{-1/3}$. Once an estimate has been made of ℓ, i.e., Eqns. (3) or (6), d may be deduced from the activation area ℓd. Since the force on the barrier is given by $F = \tau*\ell b$, these data may then be used to construct a force-activation distance (F,d) curve for the DSI[†]. The values of G_o and the (F,d) curve give a rather complete description of the DSI and may be used to test theories concerning the physical origin of the interaction.

3.   EXPERIMENTAL RESULTS: INTERSTITIAL HARDENING IN CLOSE-PACKED METALS

Octahedral sites in fcc and hcp metals are quite spacious and appreciable solid solubility of O, N, and C should be possible on the basis of size factor. Extensive solubility is observed for alloys of O and N in Ti, Zr, and Hf, which are hcp at low temperature, and for C in Th, Ni, and γFe, which have fcc crystal structures. The mechanical properties of a number of these systems have been intensively investigated, and quite detailed information is now available on the DSI in some of them. In this paper, we will compare and contrast the behavior of these close-packed interstitial alloys in order to shed light on the mechanisms involved.

3.1 HCP Alloys

(a) Hf-O. Thermally activated deformation of Hf-O alloys has recently been investigated by Das and Mitchell [18] and by Tyson and Conrad [19]. The flow stress (Fig. 5) exhibits the features described in Section 2 for ideal DBH: the low temperature strength is sensitive to T and $\dot\gamma$, and dependent on c. The temperature dependence of τ decreases at high temperatures, and above $T_c \sim 750$ K becomes approximately the same as the temperature dependence of the shear modulus C_{66} which, it should be noted, is quite large for Hf as well as for Zr and Ti. The effective stress at 0 K varies directly with $c^{\frac{1}{2}}$ (Fig. 6).

Activation energies have been calculated from temperature and strain-rate sensitivity measurements and are shown in Fig. 7. Evidence for the temperature dependence of G_o can only be obtained, at

---

[†] Note that care is required here to ensure consistency with the assumed temperature dependence of G_o, i.e., the temperature dependence of the force-distance profile.

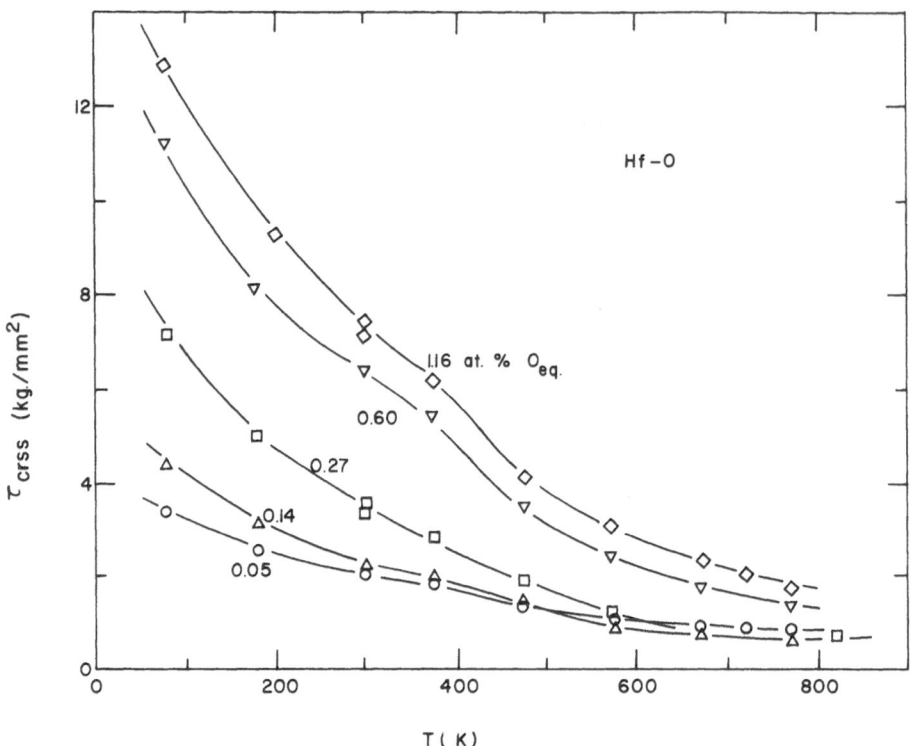

Fig. 5. Critical resolved shear stress for a series of Hf–O alloy single crystals (Tyson and Conrad [19]).

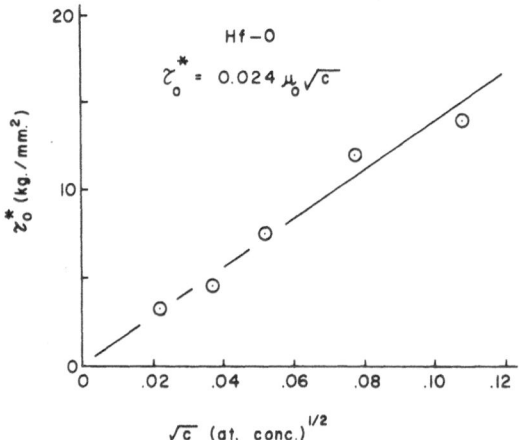

Fig. 6. Effective stress τ_o^* at 0 K for Hf-O single crystals.

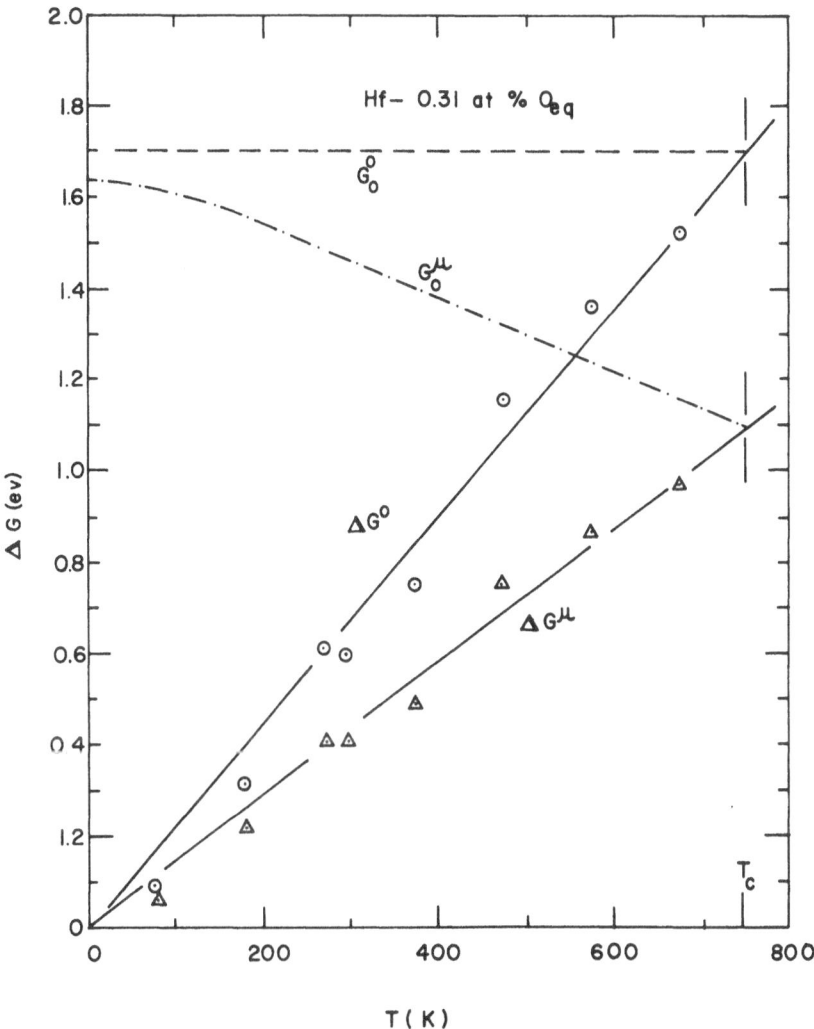

Fig. 7. Activation energies for plastic flow of Hf–O alloys, with ΔG assumed independent of temperature (ΔG^O) or proportional to the shear modulus (ΔG^μ). G_O^O and G_O^μ are total activation energies at $\tau^* = 0$ (Conrad and Tyson, Ref. [19]).

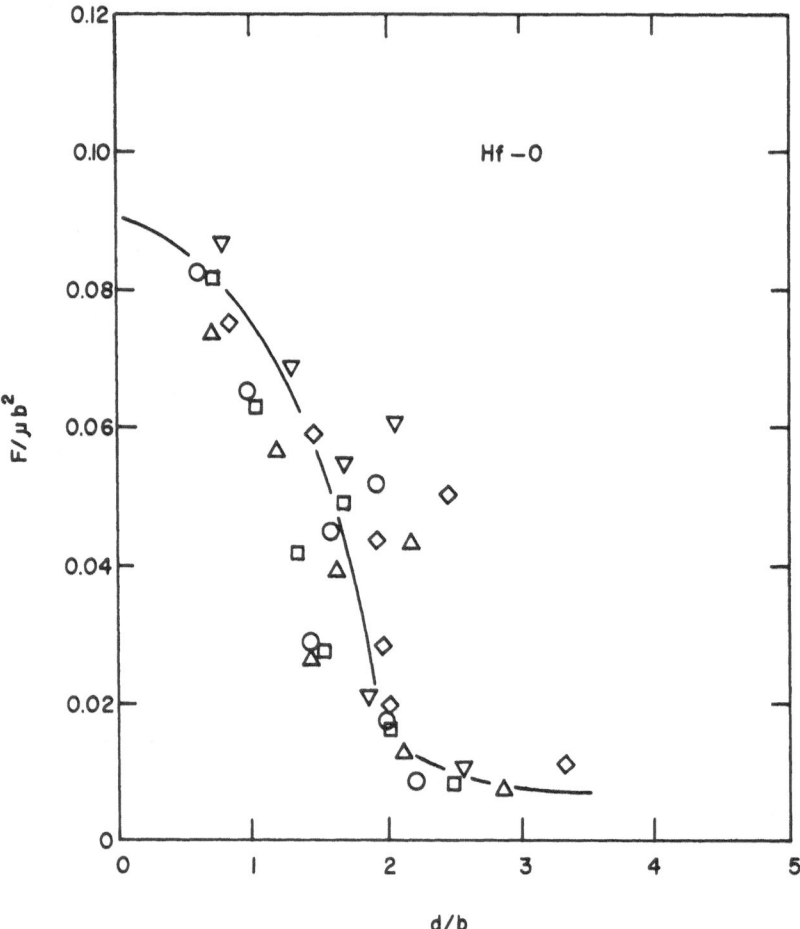

Fig. 8. Force-distance curve for DSI in Hf-O alloys (Conrad and Tyson, Ref. [19]).

the present time, on the basis of self-consistency of the thermodynamic analysis. This requires that ΔG be proportional to T. It is
evident from Fig. 7 that the degree of linearity of ΔG^o and ΔG^μ is
very similar, and so it is not clear from available data whether G_o
is temperature-independent or proportional to μ. In either case,
G_o at 0 K is $\sim$ 1.7 eV (Fig. 7).

Since $\tau_0^* \propto c^{1/2}$ (Fig. 6), it is reasonable to conclude that
dislocations interact with single interstitials. As pointed out
in Section 2, the Friedel spacing should apply in this case. From
measured values of the effective stress and activation volume, the
force-distance curve shown in Fig. 8 has been constructed by Tyson
and Conrad [19]. The range of the interaction is $\sim$ 2b and the maximum force $F_m \sim 0.09 \mu b^2$, i.e., $\alpha \sim 0.09$. The area under the curve
of Fig. 8 is approximately $0.14 \mu b^3 \sim 1.7$ eV, in agreement with G_o
of Fig. 7.

(b) Ti-O. Alloys of O in Ti have been intensively studied by
Conrad and co-workers (de Meester et al. [20]). The qualitative
features of plastic flow in this system are closely similar to those
of Hf-O alloys. Flow is thermally activated below $T_c \sim 560$ K, and
τ_0^* is again proportional to $c^{1/2}$ (Fig. 9). The slope $d(\tau^*/\mu)/dc^{1/2}$
is larger for Ti-O than for Hf-O alloys, although the critical temperature T_c for $\tau^* = 0$ is lower (560 K for Ti-O, 750 K for Hf-O).
This may be traced to a higher value of $\alpha = F_m/\mu b^2$ for O in Ti; from
Fig. 10, we find $\alpha \sim 0.21$, whereas $\alpha \sim 0.09$ for O in Hf. In con-
structing Fig. 10, the Friedel spacing (Eqn. (6)) has been assumed.
The area in Fig. 10 is $G_o \sim 1.4$ eV, in good agreement with values
calculated from thermodynamic analysis of de Meester et al. [20]
using the deformation partials $(\partial\tau^*/\partial T)\dot{\gamma}$ and $(\partial\tau^*/\partial\ln\dot{\gamma})_T$. G_o is
considerably smaller for Ti-O than for Hf-O alloys (Table II), and
so $T_c = G_o/(k\ln\dot{\gamma}_o/\dot{\gamma})$ is smaller in the former case although, as noted
above, $F_m/\mu b^2$ is larger for Ti-O than for Hf-O alloys.

(c) Zr-O. The mechanical properties of Zr-O alloys are qualitatively similar to Ti-O and Hf-O alloys, as expected from the similarities in the physical properties of the (Ti, Zr, Hf) series. Flow
is thermally activated below $T_c \sim 550$ K with an activation energy
$G_o \sim 1.4$ eV. However, τ_0^* seems to vary linearly with c rather than
$c^{1/2}$ (Fig. 11). This does not fit the single-barrier theories, and
for this reason no attempt has been made to derive a force-distance
profile for this system.

(d) Zr-N. Values of T_c and G_o for Zr-N alloys are reported
in Table II; again, the behavior is qualitatively similar to that
of the alloys discussed above. According to Tyson [10], τ_0^* varies
linearly with $c^{1/2}$, and so it is reasonable to deduce that dislocations are activated over single solute atoms in this case.

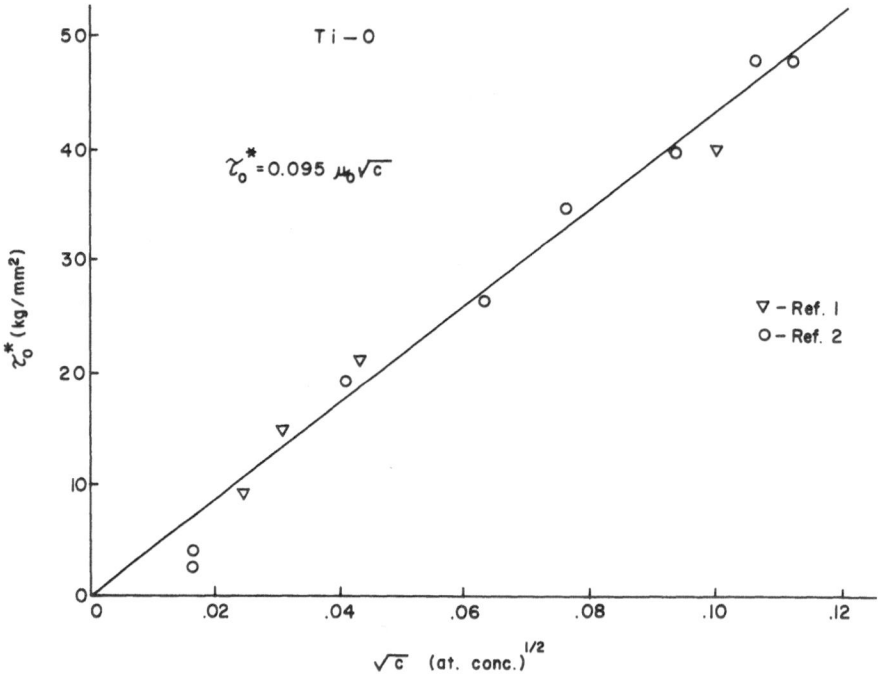

Fig. 9. Variation of effective stress at 0 K with atomic concentra-
tion c of 0 interstitials in Ti. Data from (1) H. Conrad and R.L.
Jones, in Science, Technology, and Application of Titanium, Eds. R.
Jaffee and N. Promisel, Pergamon Press (1970) p.489. (2) H. Conrad,
Acta Met. 14, 1631 (1966).

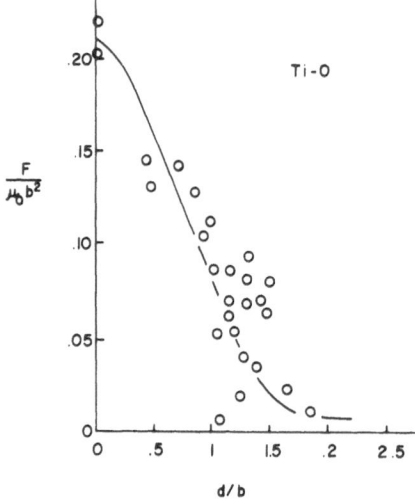

Fig. 10. Force-distance profile for 0 interstitials in Ti (Ref.
[20]).

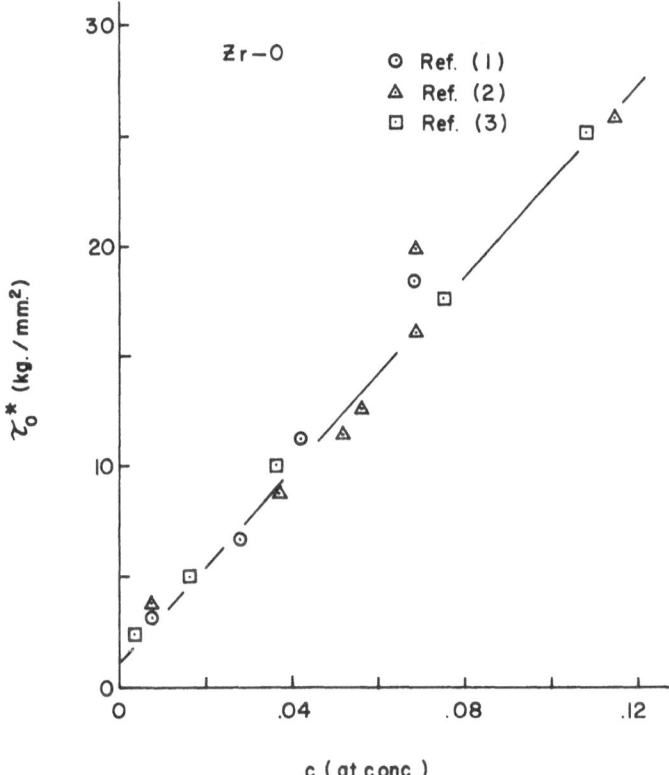

Fig. 11. Variation of τ_0^* with c for Zr-0 alloys. A Taylor factor of M = σ/τ = 2.5 has been used to convert polycrystal tensile stresses in Heaslip's work to shear stresses. Data from (1) D. Mills and G.B. Craig, Trans. TMS-AIME <u>242</u> 1881 (1968), (2) P. Soo and G.T. Higgins, Acta Met. <u>16</u> 177 (1968), and (3) T.W. Heaslip, M.A. Sc. thesis, University of Toronto (1967).

3.2 FCC Alloys

(a) <u>Ni-C</u>. Appreciable amounts of C can dissolve interstitially
in Ni and cause significant strengthening at low temperatures. According to Zwell, <u>et al.</u> [21], the change in lattice parameter is
accurately linear with concentration c, indicating that C dissolves
almost entirely as single interstitials, although there is some
evidence from internal friction measurements for the existence of
clustering. The mechanical properties of polycrystalline Ni-C
alloys have been studied thoroughly by Nakada and Keh [22]; there
is marked strengthening (Fig. 12) below 200 K, and the flow stress
is sensitive to temperature and strain rate in this region. τ_o^*
varies linearly with c (Fig. 13) in this system, and the activation
energy rises to $G_o \sim 0.5$ eV at $T_c \sim 200$ K.

The "internal stress" theory of Mott and Nabarro accounts
satisfactorily for these observations, while theories based on independent interaction with single barriers cannot explain the concentration dependence of τ_o^*. From Eqn. (10) using $\varepsilon = 0.21$, we find
$\tau_o^*/\mu = 0.312$ c, which is within a factor of two of the experimental
results of Fig. 13. This good agreement is a strong indication that
the Mott-Nabarro theory is applicable in this system, and that dislocations move over groups of solute atoms in the activation process.
Within the framework of the Labusch theory, taking reasonable values
$\alpha \sim 0.06$, $c/b^2 \sim 0.02$, $w/b \sim 1.5$, we find $\beta \sim 1$ and so dislocations
should interact with several obstacles. However, this value of β
is at the lower limit of the $c^{2/3}$ region, and a transition to the
Mott-Nabarro result of $\tau_o^* \propto c$ should not be excluded until theory
or computer simulation clarifies the nature of the flow process for
values of $\beta \lesssim 1$.

Table II. Total Activation Energies G_o for Thermal Activation and
Critical Temperature T_c for $\tau^* = 0$ for Interstitial Solutes in
Close-Packed Metals

System	Crystal Structure	T_c (K)	G_o (eV)
Zr-O	hcp	550	1.4
-N	hcp	630	1.4
Ti-O	hcp	560	1.4
-N	hcp	---	---
Hf-O	hcp	750	1.7
Ni-C	fcc	200	0.5
Th-C	fcc	350	0.5-0.9

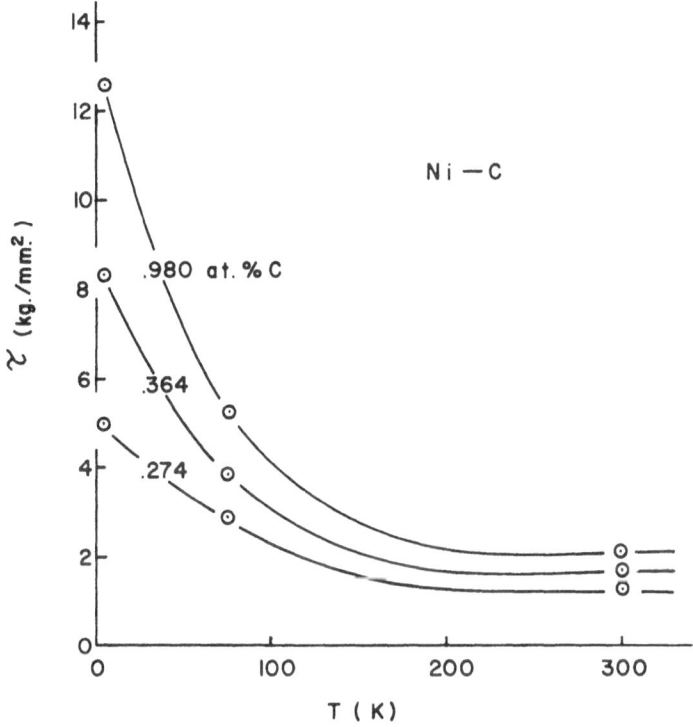

Fig. 12. Friction stress (yield stress extrapolated to infinite grain size) for Ni-C alloys (Nakada and Keh, Ref. [22]). A Taylor factor of M = 3.06 has been used to convert polycrystal tensile stress to shear stress.

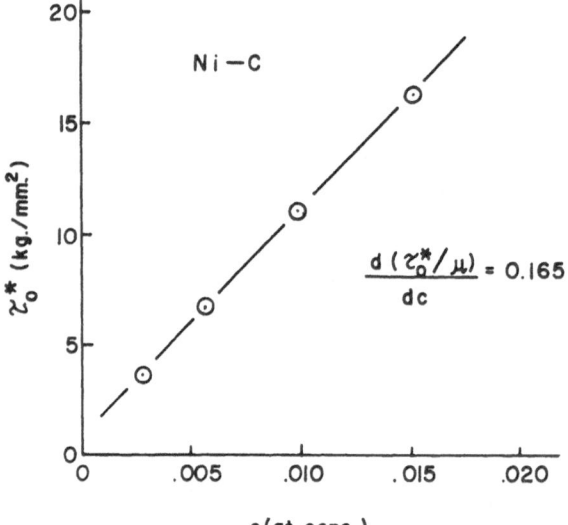

Fig. 13. Variation of effective stress τ_o^* at 0 K with concentration
of C in Ni (Nakada and Keh, Ref. [22]).

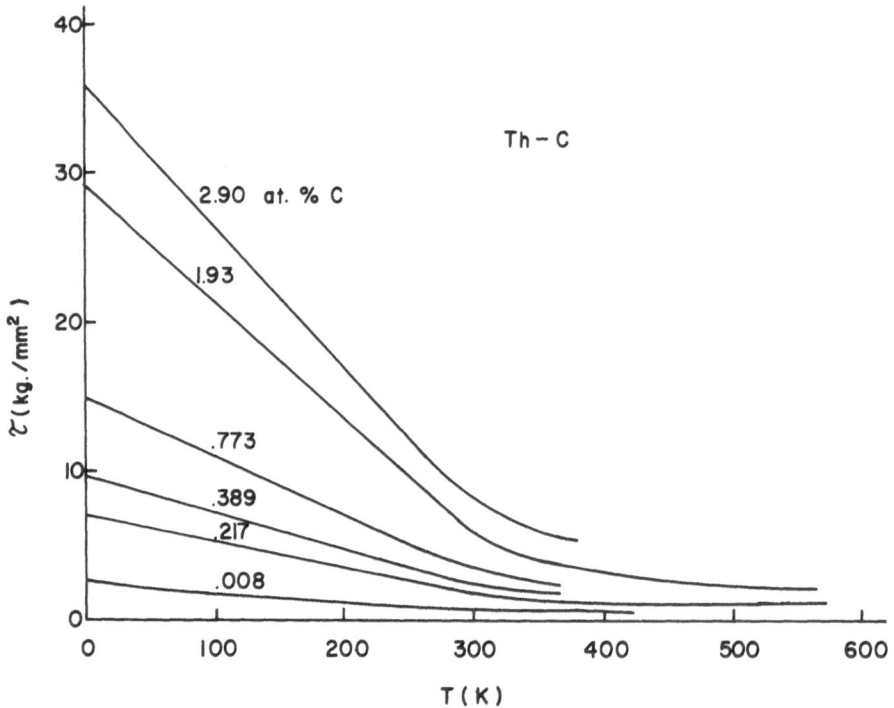

Fig. 14. Yield stress at 0.1% offset using a Taylor factor M = 3.06
for Th-C polycrystalline alloys (Peterson and Skaggs, Ref. [23]).

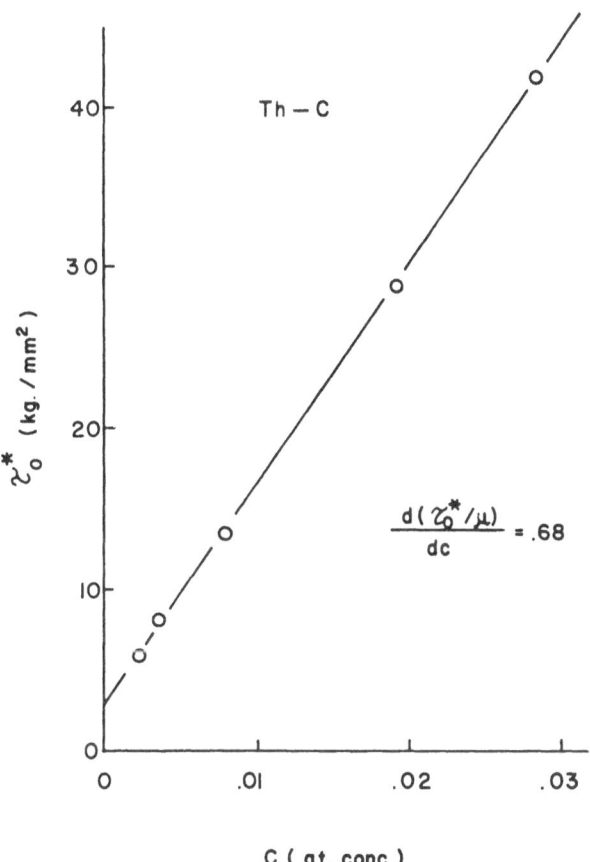

Fig. 15. Variation of τ_o^* with concentration for Th-C alloys.

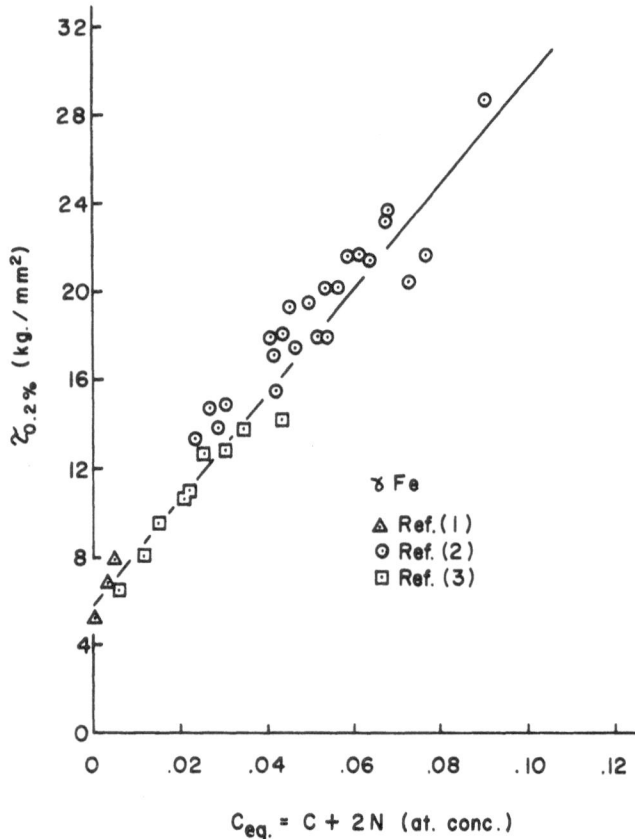

Fig. 16. Room temperature yield stress at 0.2% offset using M = 3.06 for polycrystalline interstitial alloys of austenite. The "carbon equivalent" has been taken as the sum of the C concentration plus two times the N concentration. Data from (1) D.L. Douglas, _et al._, Corrosion 20 15 (1964); (2) A. Kasak, _et al._, Proc. ASTM 59 786 (1959); and (3) K.J. Irvine, _et al._, J. Iron and Steel Inst., Oct. 1961, p 153.

(b) <u>Th-C</u>. C dissolves extensively in Th and causes appreciable strengthening below $T_c \sim 350$ K (Fig. 14). As in the case of Ni-C alloys, τ_0^* varies linearly with c (Fig. 15). However, the internal stress theory of Mott and Nabarro predicts $\tau_0^*/\mu = 0.133c$ using $\varepsilon = 0.111$, much less than the observed value of 0.68c. It is apparent in this case that the strengthening effect is more than expected from the internal stresses caused by size misfit. The experimental activation energy for flow is at least 0.5 eV, possibly as high as 0.9 eV (Peterson and Skaggs [23]).

(c) <u>γFe-(C+N)</u>. The room temperature yield strength of austenitic steels is sensitive to the interstitial content as shown in Fig. 16. There seems to be a linear variation of flow stress with C equivalent, defined as $C_{eq} = C + 2N$ since N has twice the effect of C according to the data of Irvine, <u>et al.</u> [24]. However, more data are needed on the temperature and strain rate dependence of the flow stress to aid in identification of the mechanism involved. Calculations of the DSI energy presented below indicate that the elastic size effect is large.

4. DISLOCATION-SOLUTE INTERACTION ENERGIES

In all of the alloys discussed in Section 3, there is a strong interaction between dislocations and interstitial solute atoms. Interaction energies vary between 0.5 and 1.7 eV (Table II), and $\alpha = F_m/\mu b^2$ is of the order of 0.2 (Fig. 10). In this section, possible physical mechanisms for the DSI will be examined. Such mechanisms have received considerable attention in the literature, and a number of fine reviews are available (Cottrell, [25]; Fleischer, [6]; Haasen, [26,27]). No attempt will be made at an exhaustive survey, but the principal features of mechanisms which should be relevant to the alloys under discussion will be outlined.

Mechanisms of DSI are normally divided into three classes: elastic, chemical, and electrical. These may be further divided into short-range and long-range components, although this division is somewhat arbitrary; the relevant parameter for DBH is usually the range as a fraction of the interparticle spacing.

4.1 Elastic Interactions

If a foreign atom is inserted in a lattice in the presence of a stress field from a dislocation, work is done against the stresses if the foreign atom (a) distorts the lattice, or (b) locally alters its elastic properties.

(a) The first of these effects produces an energy given by

$$G^{DS} = \sigma_{ik}^D \, \varepsilon_{ik}^S \, V \quad ,$$

Table III. DSI Energies Calculated from Size Misfit Interaction

System	$\varepsilon_1 = \dfrac{1}{a}\dfrac{da}{dc}$	$\varepsilon_3 = \dfrac{1}{c'}\dfrac{dc'}{dc}$	μ $(10^{12}\,d/cm^2)$	b (Å)	G^{DS} (eV)
Zr–O	0.039[1]	0.040[1]	0.441[5]	3.23	0.15
Zr–N	0.030[1]	0.050[1]			0.13
Ti–O	0.008[1]	0.135[1]	0.446[5]	2.95	0.11
Ti–N	0.027[1]	0.167[1]			0.17
Hf–O	0.026[1]	0.060[1]	0.578[5]	3.19	0.16
Th–C	0.111[2]		0.195[6]	3.59	0.26
Ni–C	0.21[3]		0.68[6]	2.49	0.58
γFe–C	0.20[4]		0.48[7]	2.53	0.40
γFe–N	0.21[4]		0.48[7]	2.53	0.42

References:

(1) Sources for ε_1 and ε_3 for Zr, Ti, and Hf are listed by W.R.
     Tyson, Can. Met. Quart., 6 301 (1968), where a discussion of
     the size effect interaction for a/3 <11$\bar{2}$0> ($\bar{1}$100) dislocations
     may also be found. Note that ε_1 and ε_3 in Table III are twice
     those given by Tyson. The present values are appropriate
     for extrapolation to an atomic ratio of 1:1 and a volume per
     interstitial equal to the atomic volume, rather than to an
     atomic concentration c = 0.5 and volume per solute $\sim a^2c'/2$ as
     used previously; the difference in G^{DS} using these different
     approximations is small. For discussions of this problem, see
     J.P. Hirth and M. Cohen, Scripta Met. 4 167 (1970).
(2) D.T. Peterson and R.L. Skaggs, Trans. TMS-AIME 242 922 (1968).
(3) L. Zwell, et al., Trans. TMS-AIME 242 765 (1968).
(4) N. Ridley, et al., Trans. TMS-AIME 245 1834 (1969).
(5) E.S. Fisher and C.J. Renken, Phys. Rev. 135 A482 (1964).
(6) G. Simmons, J. Grad. Res. Centre, S.M.U., Dallas, Texas, 34
     1 (1965).
(7) M.C. Mangalick, N.F. Fiore, Trans. AIME 242 2363 (1968).

Calculations:

G^{DS} is given for edge dislocations in close-packed metals by

hcp $\dfrac{a}{2}$ <11$\bar{2}$0> (1$\bar{1}$00) $G^{DS} = \dfrac{2DV}{r}(\varepsilon_1 + \varepsilon_3\nu)$

        where $V = \dfrac{\sqrt{3}}{4}a^2c'$

and a, c' are lattice parameters.

Table III. DSI Energies Calculated from Size Misfit Interaction, Continued

$$\underline{fcc} \qquad \frac{a}{2} <110> \quad (1\bar{1}1) \qquad\qquad G^{DS} = \frac{2DV}{r} (1 + \nu)\varepsilon_1$$

where $V = b^3/\sqrt{2}$.

In both equations for G^{DS}, $D = \dfrac{\mu b}{2\pi (1 - \nu)}$ and r has been taken as $\sim b$ to obtain an estimate of G_0^{DS}. The shear modulus μ has been calculated from single-crystal elastic constants as the modulus for shear on the slip plane in the slip direction; $\mu = c_{66}$ for $<11\bar{2}0>$ $(1\bar{1}00)$ (hcp), $\mu = 3c_{44}(c_{11} - c_{12})/(4c_{44} + c_{11} - c_{12})$ for $<110>$ $(1\bar{1}1)$ (fcc). In all cases, ν has been taken as 0.3.

where summation over repeated indices is implied; σ_{ik}^D is the dislocation stress field and ε_{ik}^S is the strain caused by the solute within the volume V. This formula reduces to those given in Table III, in the approximation of elastic isotropy for interstitials in octahedral sites interacting with $a/3 <11\bar{2}0>$ edge dislocations on prism planes in hcp metals or $a/2 <110>$ edge dislocations on close-packed planes in fcc metals. Corresponding values of G^{DS} calculated for the alloy systems considered above are listed in the Table. This "size misfit interaction" is much smaller for screw dislocations, but does exist due to second-order effects which can cause dilatation around a screw dislocation [28].

(b) The second elastic effect, termed the "modulus interaction", varies more rapidly with distance than the size interaction and can be significant, at least in fcc substitutional alloys [27]. It has the following form, for screw dislocations [28]

$$G^{DS} = \mu b^2 \eta \cdot V/(8\pi^2 r^2) \quad,$$

where

$$\eta = \frac{1}{\mu} \frac{d\mu}{dc}$$

V = atomic volume

r = solute-dislocation separation.

Few measurements have been made of the effect of interstitials on the moduli of close-packed metals, but taking $\eta \sim 1$ as a reasonable upper limit gives (with $V \sim b^3$ and $r \sim b$) $G^{DS} \sim 0.01~\mu b^3$, compared with values of $\sim 0.03~\mu b^3$ for the size interaction listed in Table III.

4.2 Electrical Interactions

All DSI effects are electrical in a sense, as they ultimately
can be traced to the electrostatic perturbation of the electronic
ground state by a solute atom. For example, overlap of core electrons contributes to lattice expansion (the "size effect"), and
alterations in the density of conduction electrons can change the
elastic properties of the crystal. However, a local charge on the
solute atom can be associated with a specifically electrical interaction with a dislocation that itself carries a charge (as can occur
in an ionic crystal) or an electric dipole (as shown to be possible
by Cottrell, et al. [29] in metals). The size of this interaction
is typically $\sim$0.05 eV for Cu alloys [25]. More work is required to
evaluate the magnitude of the interaction for interstitial alloys,
but it is reasonable to argue from the example of Cu alloys that
the effect will be small compared to elastic effects.

4.3 Chemical Interactions

The free energy of an atom in solid solution depends on the
configuration of the host atoms surrounding the solute, i.e., the
crystal structure. A stacking fault is essentially a region of different crystal structure than the matrix, and the free energy of a
solute atom in the fault should therefore be different from that of
a solute atom in the matrix. There will be a tendency for segregation between stacking fault and matrix to occur. According to
Suzuki [30], this effect could lead to pinning of dislocations which
are split into partials separated by a stacking fault.

While this mechanism was first applied to metals of low stacking
fault energy, the basic idea should be relevant even for narrow dislocations. The core region of any dislocation has a structure different from that of the matrix and so the free energy of a solute
atom must vary as it enters and leaves the core when the dislocation
moves past it on the slip plane [31]. In the case of the hcp metals
considered in this paper, slip dislocations on the prism plane are
thought to have core regions that are bcc in structure [32]. The
energy of interaction due to this "chemical effect" could therefore
be estimated roughly from the difference in free energy of solution
in the bcc phase and in the hcp phase (excluding configurational
free energy). Although the required thermodynamic data are not yet
available, from phase diagrams it is known that interstitial O and
N stabilize the hcp phase and constrict the range of stability of
the bcc phase. Hence it may be surmised that the energy of an interstitial in bcc Zr, Ti, or Hf is higher than in the hcp phase and so
the energy of the solute would be higher in the dislocation core than
far from it, which would lead to a short-range repulsive interaction.

This effect may be stated in another way as a change in the energy of the bonds between solute and solvent atoms as a dislocation moves over the slip plane. If the electronic structure in the vicinity of a solute atom can be described in terms of local covalent bonds, then a change in the number of neighbors in a faulted region (i.e., a dislocation core) would cause an energy change due to breaking or making of bonds. This effect has been demonstrated by Tyson [33] in computer simulation of DSI in a two-dimensional lattice. A rough estimate of the energy per bond can be obtained from the heat of formation of a compound from individual atoms, assuming this energy to be the sum of the energies of localized covalent bonds in the compound. For example, TiO crystallizes in the NaCl structure, each O atom being surrounded by six Ti atoms. If we assume the heat ΔH of the reaction

$$Ti(g) + O(g) \rightarrow TiO(s)$$

to be the energy of the $6N_O$ bonds per mole formed between Ti-O neighbors, we can obtain an estimate of the covalent bond energy ε. This will be an upper limit, of course, as the ionic component of the formation energy is ignored. However, it should be relevant to the bonding of interstitial O in hcp Ti, since the O has six Ti neighbors in both TiO and in the octahedral interstice in hcp Ti. Values obtained in this way are listed in Table IV. For the Ti-O bond, $\varepsilon \sim 2.1$ eV; from the formation energy of TiO_2, Sargent and Conrad [34] obtain ~ 1.4 eV for the energy per bond.

Table IV. Heats of Formation of ΔH and Bond Energies ε for Compounds with the NaCl Crystal Structure, and Heats of Sublimation ΔH_S and Dissociation ΔH_D.

Compound	$-\Delta H$ (kcal/mole)	ε (eV)
ZrN	87[1]	2.5
ZrC	48[1]	2.7
TiO	124[1]	2.1
TiN	81[1]	2.2
TiC	44[1]	2.4
HfN	88[1]	2.5
HfC	52[1]	2.7
ThC	39[2]	2.5

Table IV. Heats of Formation ΔH and Bond Energies ..., Continued.

Process	ΔH_s (kcal/mole)	ΔH_D (kcal/mole)
$Zr(s) \rightarrow Zr(g)$	146 [3]	
$Ti(s) \rightarrow Ti(g)$	113 [3]	
$Hf(s) \rightarrow Hf(g)$	150 [3]	
$Th(s) \rightarrow Th(g)$	137 [3]	
$C(s) \rightarrow C(g)$	171.7 [3]	
$1/2\ O_2 \rightarrow O$		59.2 [4]
$1/2\ N_2 \rightarrow N$		113.0 [4]

(1) H.L. Schick, Thermodynamics of Certain Refractory Compounds,
     Academic Press, New York, 1966.
(2) D.T. Peterson and R.L. Skaggs, Trans. TMS-AIME 242 922 (1968).
(3) L. Brewer, in High Strength Materials, V.F. Zackay, ed., Wiley,
     1965, p 12.
(4) L. Pauling, The Nature of the Chemical Bond, Cornell Univ. Press,
     1960, p 86.

Note: Bond energies have been calculated from

$$\varepsilon = \frac{1}{6}(-\Delta H + \Delta H_s + \Delta H_D).$$

5. DISCUSSION

5.1 HCP Alloys

     Solution hardening of Zr, Ti, and Hf by O and N can be explained
by a short-range DSI, with dislocations interacting with separate
barriers. The spacing is governed by the Friedel relation, since
$\tau_0^* \propto \sqrt{c}$ (with the exception of Zr-O alloys). Interaction energies
are large, ~1.5 eV (Table II), and most of the interaction is confined to a short range (Figs. 8 and 10). The DSI energy is too
large to be explained by elastic interactions (Table III and Ref.
[10]), and is more likely associated with chemical effects, i.e.,
localized bonding between solute and solvent atoms. Bond energies
estimated in Table IV are of the right order to explain the observed
DSI energies.

5.2 FCC Alloys

In Ni-C alloys, $\tau_o^* \propto c$ (Fig. 12) and $G_o \sim 0.5$ eV (Table II). There is a large elastic effect in these alloys (Table III), and no NiC compound exists indicating an absence of strong chemical interactions. The long range of the elastic interaction is consistent with application of the Mott-Nabarro "internal stress" theory to this system, which explains the linear concentration dependence and the magnitude of the strengthening effect.

Th-C alloys, on the other hand, seem to occupy an intermediate position. It is observed that $\tau_o^* \propto c$ (Fig. 16), but the strengthening is larger than can be explained by the Mott-Nabarro theory. There is a significant long-range elastic interaction (Table III), although the elastic interaction energy[†] is at least a factor of two smaller than the total DSI energy. Chemical effects should be significant, since bond energy is ~ 2.5 eV (Table IV). For these alloys, some combination of "internal stress" and "short-range interaction" theories would seem to be required. Progress on formulation of such a theory could be made by computer simulation of dislocation motion through an array of obstacles taking into account their long-range stress fields, to study in particular the effects of dislocation curvature and flexibility on the length ℓ involved in overcoming the barriers.

No attempt has been made in this paper to explain the "plateau hardening" that occurs at $T > T_c$. There is some evidence, at least in the hcp alloys studied here, that hardening above T_c can be identified primarily with internal stresses of range too large to be thermally activated, particularly if the material has been work hardened at a lower temperature. This is consistent with the technique that has been used to extract $\tau^* = \tau - \tau_\mu$ by back extrapolation from the plateau region. However, Haasen [27] has argued that plateau hardening is due primarily to elastic interactions between solutes and dislocations, at least in fcc substitutional alloys, and that low-temperature hardening for $T < T_c$ is due to short-range "core effects". This would seem inconsistent with the conclusion above that Mott-Nabarro hardening can account for the low-temperature strength of Ni-C alloys. It is possible that some form of the Mott-Nabarro theory could apply in the plateau region, with the internal stresses acting as a frictional drag on dislocations moving relatively rigidly across the slip plane.

---

† Note that Peterson and Skaggs [23] calculated a much smaller interaction energy, 0.02 eV rather than the 0.26 eV of Table III. For the volume of the strained region, they used the size of the interstitial hole rather than the atomic volume. See Hirth and Cohen, Scripta Met. 4 167 (1970) for a discussion of this problem.

6. CONCLUSIONS

Interstitial solutes in close-packed metals can cause significant solution strengthening, particularly at low temperatures. Strengthening can be due to elastic interaction or chemical effects, or to a combination of these, depending on the alloy system. In alloys with short-range chemical interactions, i.e., Zr, Ti, or Hf containing interstitial O or N, dislocations move by overcoming individual solute atoms and the Friedel spacing applies. If the barriers possess long-range elastic stress fields which account for the strengthening, such as in Ni-C alloys, the local curvature of dislocations in the internal stress field must be taken into account, following the general approach of the Mott-Nabarro theory, and barriers are overcome in groups.

REFERENCES

1. CONRAD, H., Mat. Sci. Eng. 6 265 (1970).
2. DORN, J.E., "Low-Temperature Dislocation Mechanisms", in Dislocation Dynamics, ROSENFIELD, A.R., et al., eds., McGraw-Hill, 1968, p 27.
3. DeMEESTER, B., YIN, C., DONER, M. and CONRAD, H., "Thermally Activated Deformation of Crystalline Solids", John E. Dorn Memorial Symposium, on Rate Processes in Plastic Deformation, Cleveland, Ohio, October, 1972.
4. SUREK, T., LUTON, M.J. and JONAS, J.J., Phil. Mag. 27 425 (1973).
5. BROWN, L.M. and HAM, R.K., "Dislocation-Particle Interactions", in Strengthening Methods in Crystals, KELLY, A. and NICHOLSON, R.B., eds., Elsevier, 1971, p 12.
6. FLEISCHER, R.L., "Solid Solution Hardening", in The Strength of Metals, Reinhold Press, 1962.
7. FRIEDEL, J., Dislocations, Pergamon Press, 1964, p 224
8. KOCKS, U.F., Can. J. Phys. 45 737 (1967).
9. FOREMAN, A.J.E. and MAKIN, M.J., Phil. Mag. 14 911 (1966).
10. TYSON, W.R., Can. Met. Quart. 6 301 (1968).
11. ONO, K., J. Appl. Phys. 39 1803 (1968).
12. FRIEDEL, J., Dislocations, Pergamon Press, 1964, p 380.
13. COTTRELL, A.H., Dislocations and Plastic Flow in Crystals, Clarendon Press, Oxford, 1953, p 125.
14. LABUSCH, R., phys. stat. sol. 41 659 (1970).
15. LABUSCH, R., Acta Met. 20 917 (1972).
16. NABARRO, F.R.N., J. Less-Common Metals 28 257 (1972).
17. ONO, K., "Computor Simulation of Solid-Solution Hardening", Proc. of the Third International Conf. on the Strength of Metals and Alloys, Cambridge, England, August, 1973.
18. DAS, G. and MITCHELL, T.E., Met. Trans. 4 1405 (1973).
19. TYSON, W.R. and CONRAD, H., Met. Trans. 4 2605 (1973).

20. DeMEESTER, B., DONER, M. and CONRAD, H., "Thermally Activated Deformation of Titanium at Low Temperatures", John E. Dorn Memorial Symposium, on Rate Processes in Plastic Deformation, Cleveland, Ohio, October, 1972.

21. ZWELL, L., FASISKA, E.J., NAKADA, Y. and KEH, A.S., Trans. TMS-AIME 242 765 (1968).

22. NAKADA, Y. and KEH, A.S., Met. Trans. 2 441 (1971).

23. PETERSON, D.T. and SKAGGS, R.L., Trans. TMS-AIME 242 922 (1968).

24. IRVINE, K.J., LLEWELLYN, D.T. and PICKERING, F.B., J. Iron and Steel Inst., Oct. 1961, p 153.

25. COTTRELL, A.H., "Interactions of Dislocations and Solute Atoms", Relation of Properties to Microstructure, ASM, Cleveland, 1954.

26. HAASEN, P., "Our Present Understanding of the Strength of Solid Solutions", in Alloying Behaviour and Effects in Concentrated Solid Solutions, MASSALSKI, T.B., ed., Gordon and Breach, 1965.

27. HAASEN, P., "Solid Solution Hardening", Trans. Jap. Inst. Metals (Suppl) 8 40 (1968).

28. SAXL, I., Czech J. Phys. B 14 381 (1964).

29. COTTRELL, A.H., HUNTER, S.C. and NABARRO, F.R.N., Phil. Mag. 44 1064 (1953).

30. SUZUKI, H., "Yield Strength of Binary Alloys", in Dislocations and Mechanical Properties of Crystals, FISHER, J., et al., eds., Wiley, 1957, p 172.

31. GILMAN, J.J., Micromechanics of Flow in Solids, McGraw-Hill, 1969, p 137.

32. REGNIER, P. and DUPOUY, J.M., phys. stat. sol. 28 K55 (1968).

33. TYSON, W.R., Scripta Met. 3 917 (1969).

34. SARGENT, G.A. and CONRAD, H., Scripta Met. 6 1099 (1972).

INTERATOMIC POTENTIALS AND DEFECT ENERGETICS

IN DILUTE ALLOYS

Paul S. Ho

IBM Thomas J. Watson Research Center
Yorktown Heights, New York   10598

Roy Benedek

Department of Materials Science and Engineering
Cornell University, Ithaca, New York 14850

ABSTRACT

Effective interatomic potentials for impurities in Al have been
constructed according to the pseudopotential theory. Based on a
local model potential, impurity valence and size factors have been
defined and their effects on the potential are discussed. Using
these potentials, detailed calculations based on a Green's function
lattice statics method have been made for the impurity-vacancy binding energy and for the difference in diffusion activation energies
for an impurity and a host atom. The results can account satisfactorily for the experimental data of nontransition impurities, but
less so for the noble-metal impurities. Contributions from the
lattice relaxation energies are important, particularly for impurity
migration. Dielectric screening of the ion by the conduction electrons is important in determining the potential and must be properly
accounted for in calculating the energetics for impurities.

1. INTRODUCTION

Atomic diffusion in dilute alloys is strongly influenced by the
nature of the interaction between vacancies and solute atoms. This
diffusion plays an important role in many solid state phenomena. The
formation of precipitates which contribute to alloy strengthening is
one example having considerable practical importance [1]. The phenomenon of impurity segregation to grain boundaries and voids is

79

another example. It is therefore of some interest to obtain a quantitative understanding of the vacancy-impurity interaction. The
physical parameters of primary interest in this regard are the binding energy for vacancy-solute pairs and jump frequencies for vacancy
migration in the vicinity of a solute atom.

The purpose of the present paper is to develop an approach for
the calculation of these parameters and to apply it to impurity diffusion in Al, for which extensive experimental data are available.
Such calculations require (1) interatomic potentials describing the
host-host and host-impurity interactions, and (2) a procedure for
including the effect of the lattice relaxation surrounding the point
defect configurations. An impurity is commonly characterized by
its valence and "size". We have therefore attempted to vary these
parameters in a systematic way to determine their effects on the
energetics of the vacancy-impurity interaction.

This paper is divided into two parts. The first part describes
the construction of the interatomic potential within the framework
of pseudopotential theory. Based on a local model potential, size
and valence factors are defined and their effect on the pair potentials is discussed. The second part deals with the calculation of
the impurity-vacancy binding energy and the difference in diffusion
activation energies for an impurity and a host atom. Here, considerable effort has been spent in the calculation of lattice relaxation energies. For this purpose, the Green's function formulation
of lattice statics [2] has been used. The results, particularly for
migration energies, show that lattice relaxation has a large effect.

Nontransition impurities in Al were chosen as the subject of
numerical calculations. This choice was motivated by the validity
of the pseudopotential approach for Al and also the existence of
some recent data for impurity diffusion and vacancy-solute binding
energies [3]. For simplicity, we have limited our study to non-
transition impurities. Transition impurities, which have been found
to have quite different diffusion behavior in Al, should be an interesting subject for future study.

2.   INTERATOMIC POTENTIALS

Interatomic potentials for Al alloys are developed here within
the framework of pseudopotential theory. In previous studies [4,5]
of vacancy-impurity interactions, interatomic potentials calculated
from somewhat less detailed models, e.g., the Fermi-Thomas and
Friedel potentials, have been used. While the validity of the pseudopotential approach to describe lattice defects has not been rigor-
ously established [6], this approach does represent a more serious
attempt at a realistic theory than those used in previous work.
Pseudopotential theory has at least two advantages over the above-

mentioned models: (1) the nonrealistic point ion model is avoided, and (2) the effect of electron gas exchange and correlation on the screening are included at least approximately.

In this section, the pseudopotential formulation of interatomic potentials is reviewed, and a discussion of the local model potential used in numerical calculations is given.

Consider a simple metal characterized by a nearly free electron valence band. Its total energy can be divided according to pseudo-potential theory into volume-dependent and structure-dependent parts [7]. The former includes the kinetic, exchange, and correlation energies for a uniform electron gas plus the first-order correction from the electron-ion interaction. This part can be written as a function of the average electron density and is essential for main-taining the lattice equilibrium. The second part depends on the detailed atomic arrangement. It consists of the direct electrostatic interaction between ions, and the indirect ion-ion interaction through the screening of the conduct-ion electrons. The latter is called the band-structure energy and for pure metals can be expressed as [8]

$$U_{BS} = \frac{N}{2\Omega} \sum_{k \neq o} \frac{k^2}{2\pi} \left| V_i(k) \right|^2 S(k) \left(\frac{1}{\epsilon_k} - 1\right) \quad,$$

where Ω is the atomic volume, N is the total number of ions, $V_i(k)$--called the strength function by Shaw [9]--is the Fourier transform of the bare model potential of the ion, $S(k)$ is the structure fac-tor, ϵ_k is the dielectric function, and the sum extends over all reciprocal lattice vectors. Combining the ion-ion coulomb potential with U_{BS} one can express the structure-dependent energy in terms of an effective potential $\phi(r)$ given by [8]

$$\phi(r) = \frac{z_i^2 e^2}{r} \left[1 - \frac{2}{\pi^2} \int_o^\infty \frac{(k^2 V_i(k))^2}{4\pi z_i e^2} \left(\frac{1}{\epsilon_k} - 1\right) \frac{\sin kr}{k} dk \right] \quad.$$

A similar derivation can be carried out for a dilute alloy and one arrives at an effective potential between A and B ions of the form [8]

$$\phi_{AB}(r) = \frac{z_a z_b e^2}{r} \left[1 - \frac{2}{\pi^2} \int_o^\infty \left(\frac{k^2 V_a(k)}{4\pi z_a e^2}\right) \left(\frac{k^2 V_b(k)}{4\pi z_b e^2}\right) \left(\frac{1}{\epsilon_k} - 1\right) \frac{\sin kr}{k} dk \right] \quad (1)$$

where the subscripts refer to the two types of ions. In the dilute
limit ε_k is simply the dielectric function for the pure host metal.

The separation of the total crystal energy into volume and structure dependent parts is very convenient for defect calculations in
metals. In fact, the parameters which we wish to calculate, i.e.,
binding energies and migration energies, may be expressed in terms
of structural energies alone, as is shown in the appendix. In the
calculation described in sections 3 and 4, we require the pair potentials $\phi_{AA}(r)$ and $\phi_{AB}(r)$. The remainder of this section is concerned with the determination of these quantities and particularly
with the dependence of $\phi_{AB}(r)$ on the size and valence of the impurity.

To construct the interatomic potentials, we require the explicit
form of the pseudopotential for the host and impurity ions. As in
previous work [10], we adopt a local potential having a simplified
Heine-Abarenkov [11] form:

$$V_i(r) = \begin{cases} -V_i & r < R_m^i \\ \\ \dfrac{-Z_i e^2}{r} & r > R_m^i \end{cases}$$

Here i stands for either an impurity or a host atom. This model
potential contains two parameters: V_i, the potential well depth,
and R_m^i, the model radius of the ion. The Fourier transform of $V_i(r)$
is

$$V_i(k) = \frac{-4\pi Z_i e^2}{\Omega k^2} \left[(1 - \alpha_i)\ \cos kR_m^i + \frac{\alpha_i}{R_m^i}\ \frac{\sin kR_m^i}{k} \right], \tag{2}$$

where $\alpha_i \equiv V_i R_m^i / Z_i e^2$. α is a measure of the well depth relative to
the electrostatic potential at R_m. Its value varies according to
the particular model used for the local potential, e.g., $\alpha = 0$ for
the Ashcroft "empty-core" potential and $\alpha = 1$ for Shaw's optimized
model potential. In the present work, α was fitted to give agreement with the elastic constants of Al.

On the basis of this expression for $V_i(k)$ one may discuss the
effect of valence and ionic size on the interatomic potential. The
term in the brackets depends only on the parameters α and R_m, while
the other term depends only on the ionic charge. One can consider
these terms to describe respectively the "size" and valence effects

on the interatomic potential. Substituting Eqn. (2) into Eqn. (1),
one observes that $\phi_{AB}(r)$ simply scales with the valence of the
impurity. Changes in the impurity size may be simulated by changes
in the model radius R_m^i while keeping the parameter α the same as
that of the host atom. The "size" factor, as defined by R_m^i in our
model, is incorporated into the bare ion model potential. Since
the interatomic potential is obtained by screening the bare ion
potential, the effect of the impurity size depends also on the extent
of the dielectric screening. This places the size effect on a framework quite different from some of the earlier models [12] which treat
primarily the elastic interaction between the defects.

In general, of course, α may also vary. However, appropriate
values of the parameters α and R_m for specific impurities are not
presently known. Therefore we attempt to ascertain in a general way
the effects of size and valence of the impurity on diffusion without
specifying explicitly which impurity is being considered.

$\phi_{AB}(r)$ for impurities with model radius 10% larger and smaller
than the host, along with $\phi_{AA}(r)$ for Al, is shown in Fig. 1. The
impurity potentials are seen to have forms similar to that of the
host-host potential and exhibit oscillations starting at the third
neighbor position. In the present calculations, the potentials
are truncated at the fifth neighbor position. The exact form of the
potential is extremely sensitive to the particular dielectric function
used, as recently emphasized by Duesbery and Taylor [13]. The importance of the dielectric screening is underlined by the fact that
even at the first neighbor position the effective interaction $\phi_{AA}(r)$
is reduced by 99.7% from the bare Coulomb interaction. The choice
of the dielectric function is therefore critical for the construction
of the interatomic potential. In the present work, the dielectric
function of Geldart and Taylor [14] has been employed. These authors
have computed electron gas exchange and correlation contributions in
such a way that the important compressibility sum rule is satisfied.
Our pseudopotential parameters for Al, $R_m = 0.71$ Å and $\alpha = 0.42$, have
been determined by fitting to the measured elastic constants. Phonon
dispersion curves and vacancy formation and migration energies calculated for pure Al on the basis of this pseudopotential are in good
agreement with experimental values.

Comparing the potentials in Fig. 1, one observes a systematic
change in the amplitude and phase of the oscillation as a function
of the model radius. The potential for the smaller ion has weaker
oscillations and a more advanced phase. This reflects a more complete screening of a smaller ion than a larger one. The overall
difference in the magnitudes of the potentials is quite small, which
again results from the completeness of the dielectric screening.

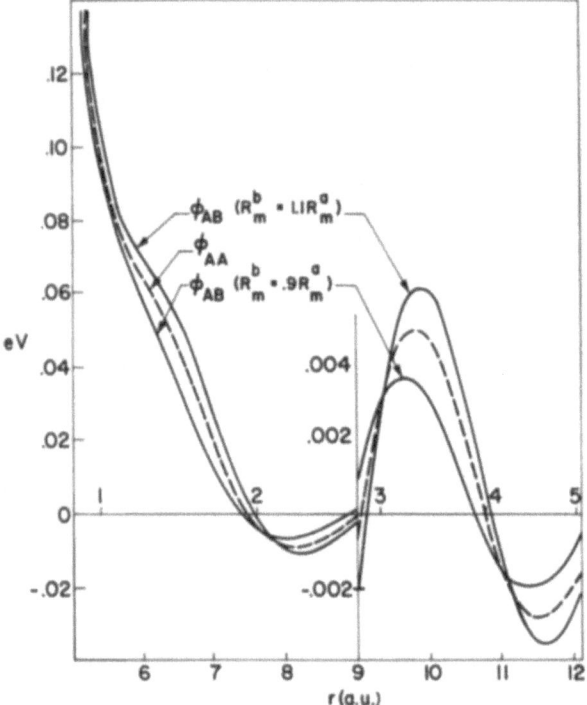

Fig. 1.   Interatomic potentials for homovalent impurities in Al.
The potential scale has been expanded tenfold for r > 9 a.u. to show
the oscillations in the potential. The positions of the first five
neighbors are also indicated. The parameters used in the host potential are R_m^a = 0.71 Å and α_a = 0.42.

3.   RELAXATION CALCULATION

In the calculation of lattice relaxation and relaxation energies,
the Green's function formulation of lattice statics [2] was employed.
In this approach, certain lattice dynamical techniques are applied in
the zero frequency limit to obtain the static equilibrium configuration of atoms around a point defect. A brief outline of the Green's
function method and how it applies to the particular defect configurations of interest in this work is given in this section.

As in the harmonic approximation of lattice dynamics, one ex-
pands the crystal energy to second order in the atomic displacements:

$$\Phi = \Phi_o - \sum_{\ell,\alpha} F_\alpha(\ell)u_\alpha(\ell) + \frac{1}{2} \sum_{\substack{\ell\ell' \\ \alpha\beta}} \phi_{\alpha\beta}(\ell,\ell')\,u_\alpha(\ell)\,u_\beta(\ell') \quad . \quad (3)$$

Here $u_\alpha(\ell) = r_\alpha(\ell) - R_\alpha(\ell)$ is the displacement of atom ℓ from its unrelaxed position (i.e., a perfect lattice site unless ℓ corresponds to an interstitial). The coefficients $F_\alpha(\ell)$ and $\phi_{\alpha\beta}(\ell,\ell')$ are determined from the interatomic potentials.

The equilibrium condition $\dfrac{\partial\phi}{\partial u_\alpha(\ell)} = 0$ may be expressed in matrix form as

$$\phi u = F$$

or

$$u = \phi^{-1}F \equiv GF \quad . \tag{4}$$

The central problem in this approach is that of inverting ϕ to obtain the Green's function G.

Any point defect may be considered to be some combination of n_v vacancies and n_i interstitials. We will refer to each particular case by the shorthand notation (n_v, n_i). For example: vacancy, $(1,0)$; interstitial $(0,1)$; "activated" or migrating vacancy, $(2,1)$; etc. The n_v vacancies and n_i interstitials comprise the defect "core".

It is convenient to divide ϕ into two parts,

$$\phi = \phi^o - \delta\phi \quad , \tag{5}$$

where $-\delta\phi$ is essentially the change in ϕ caused by the introduction of the defect. The matrix ϕ and the vectors u and F in Eqn. (4) are of dimension $D = 3(N + n_i)$ where N is the total number of lattice sites. The matrix ϕ^o takes account only of the interactions present in the perfect crystal. $-\delta\phi$ is the change in ϕ which occurs upon "turning on" the interactions of the n_i interstitial atoms and "turning off" the interactions corresponding to the n_v vacant sites. This procedure for splitting ϕ is illustrated schematically in Fig. 2 for the case of an impurity at the saddle point, i.e., $(2,1)$. For simplicity, only the (111) plane with nearest neighbor interactions is shown. In the calculation of ϕ^o the interstitial atoms are treated as noninteracting free particles. $-\delta\phi$ accounts for the interactions of the impurity atom and the "turning off" of the interactions for the vacant site at the origin.

The substitution of Eqn. (5) into Eqn. (4) yields

$$G = (1 - G^o\delta\phi)^{-1}G^o = G^o\,\delta\phi G \quad , \tag{6}$$

where $G^o \equiv (\phi^o)^{-1}$.

Eqn. (6) is the familiar Dyson equation for the Green's function. The method for solving this equation is described in detail by

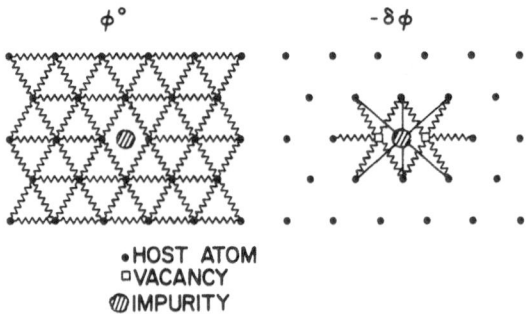

Fig. 2. Schematic illustration of splitting ϕ into ϕ^o and $-\delta\phi$ for the saddle-point configuration of an impurity ω_2 jump

Tewary [2], and therefore will only be discussed in general terms here. Two features simplify the analysis. First, the assumption of finite range interatomic interactions (in the present calculations we consider interactions extending to fifth nearest neighbors) implies that only a small number of the coefficients $F_\alpha(\ell)$ and $\delta\phi_{\alpha\beta}(\ell,\ell')$ are nonzero. By employing the procedure of matrix partitioning [15] to Eqn. (4), one obtains the relation

$$u_p = g\, f \quad , \tag{7}$$

where

$$g = g^o + g^o \delta\phi_p g \quad .$$

Eqn. (7) governs the displacements of atoms in the "perturbed space", which consists of the defect core plus the n_c lattice atoms which interact directly with the core (for the configuration shown in Fig. 2, $n_c = 116$). The matrices g, g^o, and $\delta\phi_p$, and the vectors u_p and f are of dimensions $3n_p$ where $n_p = n_i + n_v + n_c$.

Once g has been determined, displacements of atoms outside the perturbed space may also be calculated [15]. However, this is not necessary if one is only interested in the relaxation energy. Substituting Eqn. (4) into Eqn. (3), one may express the relaxation energy as

$$E_R = \phi - \phi_o = -Fu + \frac{1}{2} u\phi u = -\frac{1}{2} Fu = -\frac{1}{2} fu_p \quad . \tag{8}$$

The second simplifying feature derives from the symmetry of the lattice relaxation field. Some of the atomic displacements $u_\alpha(\ell)$ are equal to others by virtue of this symmetry and therefore the number of degrees of freedom in the perturbed space is less than $3n_p$. This allows one to obtain a "reduced" version of Eqn. (7):

$$u^r = g^r f^r \quad . \tag{9}$$

Here the reduced vectors and matrices are defined by the relations

$$u_i^r = \sum_{\ell\alpha} \psi_\alpha^i(\ell)\, u_\alpha(\ell) \quad , \tag{9a}$$

$$f_i^r = \sum_{\ell\alpha} \psi_\alpha^i(\ell) f_\alpha(\ell) \quad , \tag{9b}$$

$$g_{ij}^r = \sum_{\substack{\ell\ell' \\ \alpha\alpha'}} \psi_\alpha^i(\ell)\, g_{\alpha\beta}(\ell,\ell')\, \psi_\beta^j(\ell) \quad , \tag{9c}$$

where the ψ^i are "basis" vectors in the perturbed space. The number of distinct basis vectors having the proper symmetry is equal to n_r, the number of degrees of freedom in the perturbed space. The perturbed space displacements are related to the u_i^r by the inversion formula

$$u_\alpha(\ell) = \sum_i^{n_r} \psi_\alpha^i(\ell)\, u_i^r \quad . \tag{10}$$

The reduced Green's function is determined from the Dyson equation

$$g^r = [(1 - g^o \delta\phi)^r]^{-1}\, g^{or} \equiv (A^r)^{-1}\, g^{or} \quad . \tag{11}$$

Eqn. (8) in conjunction with Eqns. (9) – (11) provide a means for calculating relaxation energies. The defect configurations for which we require relaxation energies are illustrated in Figs. 3 and 5, and are discussed in detail in sections 4 and 5. We wish here to point out that for each configuration the defect core may be found by inspection. Migrating atoms at saddle point sites and impurity atoms are treated formally as interstitials.* A substitutional impurity is treated as an interstitial superimposed on a vacancy. For example, substitutional impurity atom, (1,1); migrating impurity atom (ω_2 jump), (2,1); dissociative jump configuration

---

* An impurity atom is treated formally as an interstitial even if it occupies a substitutional site since its force constants differ from those of the host.

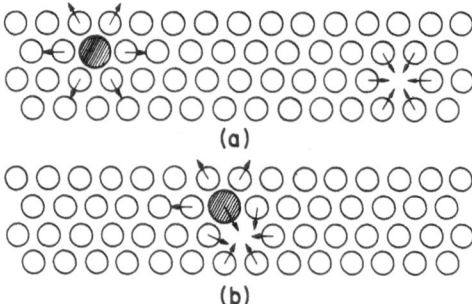

Fig. 3. The formation of an impurity-vacancy pair in (b) is accomplished by bringing together the isolated vacancy and impurity in (a).

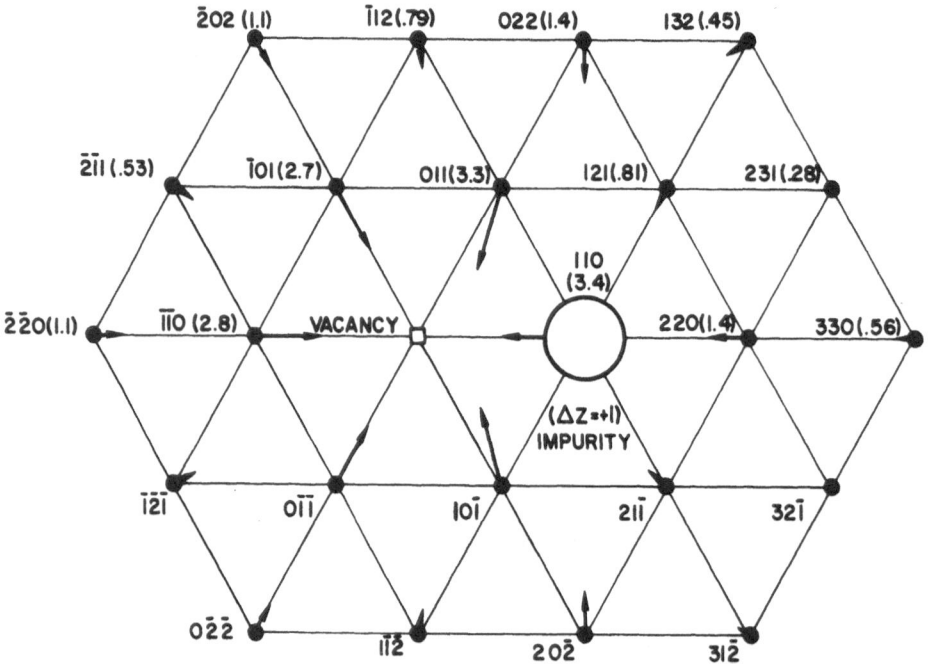

Fig. 4. (111) projection of displacement field for vacancy-impurity pair in Al. This impurity has Z_b = 4 but the same R_m as Al.

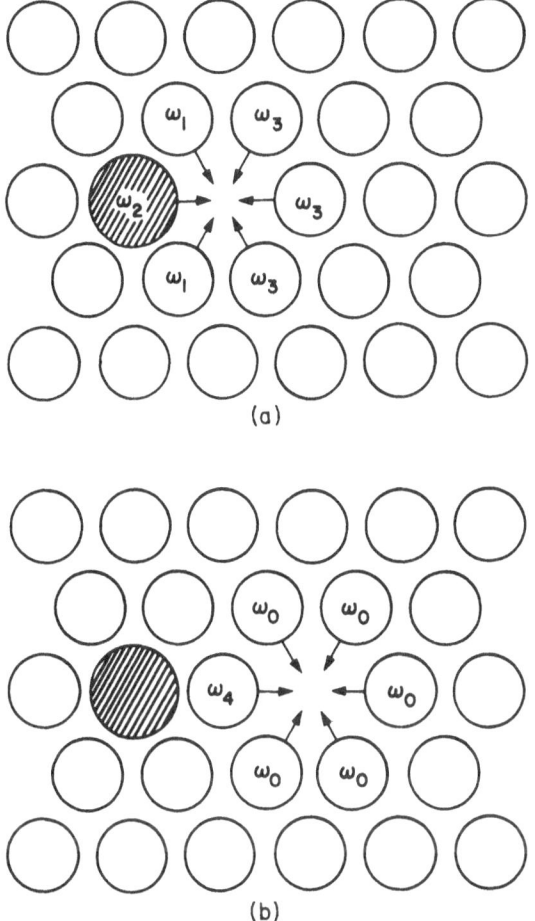

Fig. 5. Various atomic jump frequencies entering kinetic analysis
of impurity diffusion in the Howard–Manning [17] model

(ω_3 jump), (3,2); etc. Once the defect core for a particular configuration is identified and the appropriate interatomic potentials
are specified, the calculation of the relaxation energy may proceed
as we have outlined in this section.

Finally, some technical details should be mentioned here. The
dimension $3n_p$ of the perturbed space is typically large enough so
that direct multiplication to obtain the product $g^o \delta\phi$ in Eqn. (11)

is quite time consuming, even when performed on a computer. It is somewhat more convenient to compute the matrix A^r from the relation

$$A^r = \tilde{A}^r + \delta A^r \quad , \tag{12}$$

where

$$\tilde{A}^r = 1 - g^{or}\delta\phi^r \quad , \tag{12a}$$

and

$$\delta A^r_{ij} = - \sum_{\ell\alpha} \sum_{\ell'\alpha'} \sum_{\ell''\beta}^{n_v} \psi^i_\alpha(\ell) \; g^o_{\alpha\beta} \; (\ell,\ell'')\delta\phi_{\beta\alpha},(\ell'',\ell) \; \psi^i_{\alpha'}(\ell')\psi^i_{\alpha'}(\ell'). \tag{12b}$$

The reduced matrices g^{or} and $\delta\phi^r$ are defined in analogy to g^r in Eqn. (9c). In Eqn. (12b), the ℓ'' sum includes the n_v core vacancies. The basis vectors ψ possess no nonzero elements in the space of the core vacancies. Therefore, the contribution to A^r due to these vacancies is accounted for explicitly in the term δA^r. In the case of a monovacancy or a substitutional impurity, it is easy to show that $\delta A^r = 0$ as a consequence of the inversion symmetry about the defect site. For more complicated configurations, δA^r is in general nonzero and must be accounted for.

In the lattice statics calculations performed in this work, pairs of atoms were assumed to interact via central forces. In this model, the ϕ matrix may be expressed in terms of the first two derivatives of the interatomic potential:

$$\phi_{\alpha\beta}(\ell,\ell') = (-\delta_{\alpha\beta} + \frac{r_\alpha r_\beta}{r^2}) \frac{1}{r} \frac{d\phi(r)}{dr} - \frac{r_\alpha r_\beta}{r^2} \frac{d^2\phi(r)}{dr^2}, \; \ell' \neq \ell \; ,$$

$$\phi_{\alpha\beta}(\ell,\ell) = - \sum_{\ell'\neq\ell} \phi_{\alpha\beta} \; (\ell,\ell') \quad ,$$

where $\vec{r} = \vec{R}(\ell) - \vec{R}(\ell')$ and $\phi(r)$ is the interatomic potential connecting atoms ℓ and ℓ'.

4. IMPURITY-VACANCY BINDING ENERGY

Fig. 3(b) illustrates a vacancy-impurity pair in the nearest-neighbor configuration, while Fig. 3(a) shows such a pair separated

by a large distance. The binding energy E_b is the difference* in
the crystal energies for these two configurations. In the appendix
we show that this difference may be expressed in terms of pair interactions alone. It is convenient to divide the binding energy
into two parts; one, which we call $\mathbf{E_b^C}$, corresponds to the unrelaxed
lattices, while the other, E_b^R, accounts for the lattice relaxation.
Referring to Figs. 3(a) and 3(b), the "configuration" energy E_b^C may
be written down immediately:

$$E_b^C = \phi_{AB}(r_1) - \phi_{AA}(r_1) \quad,$$

where r_1 is the nearest neighbor distance. The calculation of E_b^R
is more involved since it requires the application of the lattice
statics method described in the previous section to obtain the relaxation energies for the pair, the isolated vacancy and the substitutional impurity. If these energies are called respectively
E_R^{vs}, E_R^v and E_R^s, one may write

$$E_b^R = E_R^s + E_R^v - E_R^{vs}.$$

The calculation of the first two terms is relatively easy since
$n_r = 7$, and $\delta A^r = 0$. The calculation of E_R^{vs} is more difficult,
however, since $n_r = 96$ and $\delta A^r \neq 0$. A useful self-consistency check
on the latter calculation is available. If the solute is treated
as simply another host atom in the impurity-vacancy pair calculation, one should obtain results identical to those for the isolated
vacancy. This was in fact verified in our calculations.

    In Fig. 4 is illustrated the displacement field in a (111)
plane containing a vacancy-impurity pair, for an impurity of valence
4. As required, the displacements exhibit axial symmetry about the
vacancy-impurity line. The magnitude of the individual atomic displacement reflects the extra attraction associated with the impurity;
compare, for example, the 2.8% inward displacement of the $\overline{1}\overline{1}0$ atom
with the 3.4% displacement of the impurity.

    In Table I, the results of the binding energy calculation are
given. For impurities which differ only in valence from the host,
the binding energy results primarily from E_b^C. However, the relaxation energy E_b^R is not negligible and it always reduces E_b. The sign
of E_b depends on the sign and magnitude of $\phi_{AA}(r_1)$, which in turn
depends on the dielectric screening. In the present case, since
$\phi_{AA}(r_1)$ is positive, a divalent impurity is repelled by the vacancy

---

* We adopt the usual sign convention for the binding energy; E_b is
  positive if the crystal has lower total energy in the nearest
  neighbor configuration.

and a negative E_b is expected. The opposite situation exists for a quadravalent impurity. For impurities differing only in size from the host, E_b^R is quite important relative to E_b^C, as one can see from the results in Table I. The sign of E_b for such impurities depends on the balance of E_b^C and E_b^R.

Table I. Theoretical Binding Energy for Impurity-
Vacancy Pairs in Al

(All units in eV)

Impurity	$\phi_{AB}(r_1)$	E_b^c	E_R^s	E_R^{vs}	E_b^R	E_b
$Z_b = 2$ $R_m^b = R_m^a$	.069	−.034	−.008	−.104	.012	−.022
$Z_b = 4$ $R_m^b = R_m^a$	.138	.035	.007	−.075	−.016	.019
$Z_b = 3$ $R_m^b = 0.9R_m^a$	.099	−.004	−.002	−.091	.005	.001
$Z_b = 3$ $R_m^b = 1.1R_m^a$	.115	.012	−.005	−.075	−.014	−.002

NOTE: $\phi_{AA}(r_1) = 0.103$; $E_R^v = -0.084$

Overall, the binding energies are quite small. It is possible to obtain an upper bound for the binding energy of impurities by setting E_b^c equal to $\phi_{AA}(r_1)$ and ignoring the contribution from the relaxation energy. Based on the effective potential for Al in Fig. 1, this upper limit is estimated to be 0.1 eV. This value, of course, depends on the dielectric function used in calculating the potential. However, judging from the magnitude of other Al potentials, e.g., those derived by Duesbery and Taylor [13] and Shyu, et al. [16], the value of 0.1 eV appears to be a valid estimate.

5.   ACTIVATION ENERGY FOR IMPURITY DIFFUSION

The kinetics for impurity diffusion in an fcc lattice has been studied by Howard and Manning [17]. Figs. 5(a) and 5(b) depict the various atomic jumps entering their model. The ratio of the diffusivities of the impurity and the host atom can be expressed as [17]

$$\frac{D_b}{D_a} = \left(\frac{f_b}{f_a}\right) \left(\frac{\omega_2}{\omega_0}\right) \left(\frac{\omega_4}{\omega_3}\right) \quad ,$$

where the first term is the ratio of the correlation factors, the second term is the ratio of the frequencies of the saddle-point jumps for the impurity and the host, and the last term is the ratio of the associative and dissociative jump frequencies for the vacancy-impurity pair. The second term is related to the migration energy difference and the third to the binding energy. One can write the difference in the diffusion activation energies as [18]

$$\Delta Q = \Delta E - E_b - k \frac{\partial \ln f_b}{\partial \left(\frac{1}{T}\right)} \quad , \tag{13}$$

where ΔE is the difference in migration energies for the impurity and the host atom and the last term arises from the temperature dependence of the correlation factor for the impurity. The last factor can be calculated from the temperature dependence of the jump frequencies in the kinetic model. Peterson and Rothman [4] estimated it to be very close to zero for monovalent and divalent impurities in Al, independent of whether the electrostatic model or the oscillatory potential is used. These authors also reported a preliminary measurement of this factor for Zn isotopes and found a small value of less than 0.1 eV. For simplicity, we assume this factor to be zero here.

In analogy to our treatment of the impurity binding energy, ΔE may be divided into a "configuration" contribution ΔE_c associated with the unrelaxed saddle-point configurations and a contribution ΔE_R resulting from lattice relaxation. In calculating ΔE_c or ΔE_R, one has to account not only for the different saddle-point configurations but also the different initial lattices. Thus,

$$\Delta E_R = [E_R^s (\omega_2) - E_R^{vs}] - [E_R(\omega_0) - E_R^v] \quad ,$$

where the relaxation energies are associated respectively with the impurity saddle point, impurity-vacancy pair, host-atom saddle point and isolated vacancy. A similar expression exists for ΔE_c.

In Table II, numerical values calculated for ΔE are presented. It is found that the value of ΔE_c is determined mainly by the sign

and magnitude of ϕ_{AA} at the distance from the saddle-point to one
of the four "ring" lattice sites. In the present calculation, ϕ_{AA}
at that distance is positive; therefore, a divalent impurity inter-
acts less strongly with its surrounding atoms than a host atom because of its smaller valence. This gives rise to a negative ΔE_c.
A similar situation exists for the homovalent impurity with model
radius smaller than the host. The contribution from relaxation
energies is observed to be comparable to that from ΔE_c and therefore must be carefully calculated to ensure a reliable value for
ΔE.

Table II.   Theoretical ΔE for Impurity Migration in Al

(All units in eV)

Impurity	$E_C^s (\omega_2)$	ΔE_C	$E_R^s (\omega_2)$	ΔE_R	ΔE
$Z_b = 2$ $R_m^b = R_m^a$	1.318	$-.276$	$-.243$	.209	$-.067$
$Z_b = 4$ $R_m^b = R_m^a$	2.637	.276	$-.648$	$-.225$	.051
$Z_b = 3$ $R_m^b = 0.9 R_m^a$	1.619	$-.306$	$-.279$	.160	$-.146$
$Z_b = 3$ $R_m^b = 1.1 R_m^a$	2.449	.347	$-.643$	$-.220$	.127

NOTE: $E_C^s (\omega_2)$ is the "configuration" energy for the unrelaxed impurity
saddle point. Other parameters needed for calculating ΔE are:
1.977 eV for the configuration energy of the host saddle point
and -0.432 eV for $E_R (\omega_0)$.

Table III gives results for ΔQ, neglecting the final term in
Eqn. (13). Comparing the values of ΔE and E_b, one observes that the
size and valence have a larger effect on impurity migration than on
binding energies.

Table III. Theoretical ΔQ for Impurity Diffusion in Al

(All units in eV)

Impurity	ΔE	E_b	ΔQ
$Z_b = 2$ $R_m^b = R_m^a$	-.067	-.022	-.045
$Z_b = 4$ $R_m^b = R_m^a$	.051	.019	.032
$Z_b = 3$ $R_m^b = 0.9R_m^a$	-.146	.001	-.147
$Z_b = 3$ $R_m^b = 1.1R_m^a$	.127	-.002	.129

6. DISCUSSION

It is interesting to compare the present results with available data on the binding energy and activation energy for nontransition impurities in Al. In doing so, it is important to keep in mind that, since our calculations do not specify particular impurities, the comparison is only semi-quantitative in nature. Furthermore, it is difficult to extrapolate our results to very different impurity parameters. In a recent review by Balluffi and Ho [3], it was noted that the equilibrium experiments, such as the lattice expansion and positron annihilation measurements, give binding energies systematically lower than those obtained from quenching-annealing experiments. Our results seem to support the small binding energies obtained from the equilibrium measurements. For some impurities, the agreement is even quantitative; for example, Si has a valence of 4 and an E_b of 0.03 eV [19]; Mg is divalent and its E_b was measured to be -0.1 $\pm$ 0.04 eV [20] or < 0.05 ev [21].

In Fig. 6 we list experimental values of ΔQ for three solute pairs (Cu,Ag), (Zn,Cd), and (Ga,In), together with values of R_m for

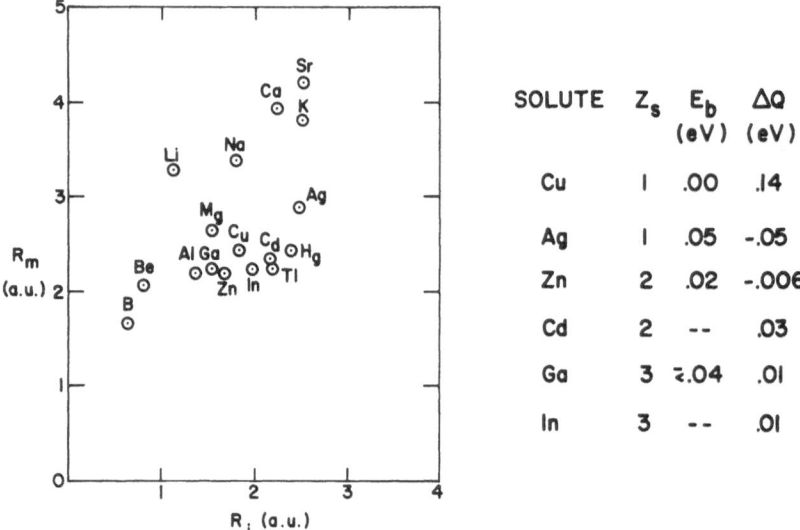

Fig. 6. Values of R_m for various elements obtained by Shaw [9] for the optimized model potential <u>versus</u> Pauling ionic radii [22], along with diffusion data for impurity in Al. The references for these data are given in Ref. [3].

various elements recently determined by Shaw [9].* In view of their model radii, the homovalent pair (Ga,In) should be almost equal in size to Al, and therefore, their ΔQ values are expected to be quite small. This is consistent with the experimental results. For the divalent pair (Zn,Cd), the valence effect is expected to dominate in Zn since its R_m is nearly equal to that of Al; on this basis, a negative ΔQ is predicted. For Cd, which is larger than Zn, the size effect is expected to compensate the valence effect. These predictions appear to be in accord with experimental data; ΔQ for Cd is larger than for Zn and has the opposite sign. However, our calculated ΔQ appear to exceed the measured values. The same line of reasoning fails when applied to the monovalent noble metal pair (Cu,Ag). Ag is larger than Cu and is therefore expected to have a higher ΔQ, contrary to experimental observation. Thus, our calculation does not seem to account even qualitatively for the behavior of noble metal impurities. This is not surprising since the applicability of a simple model potential to d-band metals is questionable.

---

* Shaw determined these R_m values based on his optimized model potention which in its local form is equivalent to setting α unity. As a result, his R_m value for Al is considerably larger than ours.

In the foregoing discussion, we have taken the model radius R_m rather than, say, its Pauling ionic radius R_i (c.f. Fig. 6) to represent the impurity "size". The former parameter is a characteristic of the metallic state and is therefore more appropriate in the present context. One may note in Fig. 6 that in certain cases a large discrepancy exists between R_m and R_i. For example, the series of trivalent elements Al, Ga, In, and Tl have almost equal model radii but their ionic radii vary by almost a factor of two.

In view of model radii plotted in Fig. 6, the monovalent pair Na and K and the divalent pair Be and Mg appear to be interesting elements for experimental study. All of these can be classified as typical simple metal ions and the model radii of the two members of each pair are quite different. At present, there are no reliable diffusion data for these impurities in Al.

Finally, we consider the validity of the almost universally adopted nearest-neighbor model for the kinetic analysis of impurity diffusion in fcc crystals. A crucial question is whether the binding energy for vacancy-impurity pair beyond the nearest-neighbor distance is negligible compared to that for a nearest-neighbor pair. To settle this point, further detailed calculation of the binding energy would be required. However, judging from the generally small values obtained here for the binding energy of nearest-neighbor pair, it seems possible that some more distant pairs can have comparable stability. If this were the case, a kinetic model incorporating more distant jumps than those shown in Fig. 5 would be required. For Al, values of the interatomic potential from the second to the fifth neighbor positions are comparable in magnitude and therefore any extension beyond the nearest-neighbor model would be very complicated.

ACKNOWLEDGMENT

One of us (R.B.) wishes to acknowledge the support provided by the U.S. Atomic Energy Commission under Contract AT (11-1)-3158.

APPENDIX

In this appendix, it is shown that the volume dependent part of the crystal energy may be neglected in binding energy and migration energy calculations.

Consider the migration of an isolated vacancy. The migration energy may be expressed as

$$E_m = \Phi_{sp}(N, \Omega_{sp}) - \Phi_v(N, \Omega_v) \quad , \tag{A.1}$$

the difference in crystal energies for the saddle point and simple
vacancy configurations. N is the number of atoms and Ω the total
volume of the relaxed crystal. Since the vacancy formation volume
differs in general from the self-diffusion activation volume,
$\Omega_{sp} \neq \Omega_v$. Subtracting and adding the energy of a perfect crystal,
$\Phi(N,\Omega)$, to Eqn. (A.1), one obtains

$$E_m = [\Phi_{sp}(N,\Omega_{sp}) - \Phi(N,\Omega)] - [\Phi_v(N,\Omega_v) - \Phi(N,\Omega)]$$

$$\equiv E_f^{sp} - E_f^v \quad . \tag{A.2}$$

The term within the second pair of brackets is simply the vacancy
formation energy. In earlier work [15], it was shown that this may
be expressed in the form

$$E_f^v = E_c^v + E_R^v - E_d - E_s \quad , \tag{A.3}$$

where the first two terms on the right hand side are, respectively,
the "configuration" and "relaxation" energies, E_d is the change in
pairwise interaction energy associated with a uniform expansion of
the crystal by one atomic volume and E_s is the pairwise interaction
energy of a surface atom. In deriving Eqn. (A.3), the perfect crystal is taken to be at equilibrium. An expression similar to Eqn.
(A.3) may be derived for E_f^{sp}:

$$E_f^{sp} = E_c^{sp} + E_R^{sp} - E_d - E_s \quad . \tag{A.4}$$

Substituting Eqn. (A.4) and (A.3) into Eqn. (A.2), one obtains

$$E_m = E_m^c + E_m^R \quad ,$$

where

$$E_m^c = E_c^{sp} - E_c^v \quad ,$$

$$E_m^R = E_R^{sp} - E_R^v \quad .$$

Thus the migration energy is expressed in terms of configuration
and relaxation contributions, both of which are determined by the
pairwise interactions.

The above discussion is concerned with the migration of an
isolated vacancy. Similar derivations may be carried out for the
impurity-vacancy binding energy and migration energies in the presence of an impurity. One obtains, for example,

$$E_b = E_b^c + E_b^R \; ,$$

a relation used in Section 4.

REFERENCES

1. See for example: Phase Stability in Metals and Alloys, RUDMAN, P.S., et al., eds., McGraw-Hill Book Co., New York, 1967.
2. TEWARY, V.K., A.E.R.E. Harwell Report T.P. 548, 1973; see also BENEDEK, R. and HO, P.S., J. Phys. F: Metal Phys. 3 1285 (1973).
3. See the recent review by BALLUFFI, R.W. and HO, P.S., in ASM Seminar for Diffusion, AARONSON, H.I., ed., Am. Society of Metals, Ohio, 1974.
4. PETERSON, N.L. and ROTHMAN, S.J., Phys. Rev. B1 3264 (1970).
5. EDELGLAR, S.M. and OHRING, M., Trans. AIME 245 186 (1969).
6. HEINE, V. and WEAIRE, D., in Solid State Physics, Vol. 24, EHRENREICH, H., et al., eds., Academic Press, 1970.
7. See for example: Pseudopotentials in the Theory of Metals, by HARRISON, W.A., Benjamin, Inc., New York, 1966.
8. ASHCROFT, N.W., in Interatomic Potentials and Simulation of Lattice Defects, GEHLEN, P.C., et al., eds., Plenum Press, New York, 1972, pp 91 to 110.
9. SHAW, R.W., Jr., Phys. Rev. B5 4742 (1972).
10. HO, P.S., Phys. Rev. B3 4035 (1971); Phys. Rev. B7 3550 (1973).
11. HEINE, V. and ABARENKOV, I., Phil. Mag. 9 451 (1964).
12. LAZARUS, D., Phys. Rev. 93 973 (1954), and SWALIN, R.A., Acta Met. 5 443 (1957).
13. DUESBERY, M.S. and TAYLOR, R., Phys. Rev. B7 2870 (1973).
14. GELDART, D.J.W. and TAYLOR, R., Can. J. Phys. 48 167 (1970).
15. BENEDEK, R. and HO, P.S., J. Phys. F: Metal Phys. 4 181 (1974).
16. SHYU, W.M., WEHLING, J.H., CORDES, M.R. and GASPARI, G.D., Phys. Rev. B4 1802 (1971).
17. HOWARD, R.E. and MANNING, J.R., Phys. Rev. 154 561 (1967).
18. LeCLAIRE, A.D., Phil. Mag. 7 141 (1962).
19. BURKE, J. and KING, A.D., Phil. Mag. 21 7 (1970).
20. BEAMAN, D.R., BALLUFFI, R.W. and SIMMONS, R.S., Phys. Rev. 137A 917 (1965).
21. McKEE, B.T.A., STEWART, A.T. and STOTT, M.J., Conf. on Point Defects and Their Aggregates in Metals, University of Sussex, England, 1972.
22. PAULING, L., The Nature of the Chemical Bond, Cornell University Press, Ithaca, New York, 1940.

POINT DEFECT-DISLOCATION INTERACTIONS

ARISING FROM NONLINEAR ELASTIC EFFECTS

R. G. Hoagland
BATTELLE
Columbus Laboratories
Columbus, Ohio 43201

J. P. Hirth
Department of Metallurgical Engineering
The Ohio State University
Columbus, Ohio 43210

and

P. C. Gehlen
BATTELLE
Columbus Laboratories
Columbus, Ohio 43201

ABSTRACT

The highly nonlinear elastic state in the vicinity of a dislocation core is the source of a linear elastic displacement field (core
field) outside the core. Recent computer simulation of dislocation
models in α-Fe and KCl have shed considerable light on the nature of
the core field and these results are reviewed. An important feature
of the core field is a net dilatation. A decrease in the elastic constants near the dislocation is also noted and results from the lattice expansion. The implications of these effects in terms of interactions of point defects with the core field is discussed.

1. INTRODUCTION

The essential physical nature of the interaction between dislocations and solute atoms can be understood in terms of isotropic
elasticity theory as developed by Eshelby [1]. For substitutional
solute atoms or interstitial atoms in sites of high symmetry there

101

is a size-effect interaction between the hydrostatic (or mean-normal)
stress component of edge or mixed dislocations and the center-of-
expansion volume strength of the solute atom. Interstitial atoms
in lower symmetry sites, such as the octahedral site in bcc metals,
interact with the full stress tensor of both edge and screw dislocations. Solute atoms in the dislocation core must be treated
separately. All other second-order effects are included in the
so-called modulus interaction, wherein the solute atom is considered
to be a small ball with bulk elastic properties of the pure solute.
Crudely, this corresponds to changes in the local atomic interaction
energy with lattice strain, but it empirically includes long-range
electronic interactions and various nonlinear elastic effects.
There are other indirect interaction effects via solute interaction
with stacking faults in extended dislocations, the Suzuki [2] effect
which has been elaborated upon by a number of researchers. One important aspect of the overall theory which is sometimes erroneously
neglected is that for concentrated atmospheres of solute atoms,
Fermi-Dirac statistics must be used to describe the interaction [3].
All of the above phenomena are adequately discussed in recent books
on dislocation theory [4-6].

     Improvements in the theory can be made by using anisotropic
elastic theory. Except for special cases in high-symmetry crystals,
all solute atoms then interact with the full stress tensor of dislocations. An indication of this is given by Masumura and Sines [7]
who treated the anisotropic interaction between a dislocation and
arrangements of point forces which are the analog of a solute atom.
Recent developments of rapid numerical techniques to evaluate certain integrals make it possible to treat the complete interaction
between solute atoms and arbitrarily curved dislocations in the
anisotropic elastic approximation by Green's function methods [8-10].
Results of such computations are not yet available, however.

     Electronic interactions must await further advances in inter-
atomic potentials for detailed analysis. However, one effect has
been established: the Fermi surface is distorted by local stress
fields leading to the redistribution of electrons and the development of an electrostatic charge [11,12].

     Nonlinear elastic effects also can influence defect interac-
tions in a number of ways. Obviously, interactions within the core
must be treated by discrete atomic interactions which are perforce
nonlinear in nature within the highly distorted core region. An indirect effect which is less obvious and which has not yet been treated
is that solutes in the core region can modify the long-range elastic
field of the dislocation, a phenomenon which is absent in the linear
elastic approximation. Other nonlinear effects are the influence
of the long-range dislocation field on the local interactions between solute atoms and an external stress field, and changes in

the interaction energy between a solute atom and a dislocation due
to an external stress field [13]. All of these nonlinear effects
require further detailed knowledge of atomic interaction potentials
before they can be quantitatively analyzed. A final nonlinear
elastic effect is that both the dislocation and the solute atom
produce a long-range linear elastic field arising from the nonlinear
elastic state at their core. This case can be developed quantitatively and is the basis for the remainder of this paper.

The quantitative treatment of the nonlinear field is given in
terms of isotropic elastic theory for simplicity in presentation,
analogous to the treatment of the Volterra dislocation interaction
discussed above. The same methods can be used in the anisotropic
case. The Green's function methods mentioned previously [8-10]
can be used to describe the long-range field arising from the dislocation core in the anisotropic case.

2. NONLINEAR ELASTIC DISLOCATION FIELDS

A simple approximation which eliminates the singularity at the
dislocation origin in continuum solids is to remove the cylindrical
region containing the dislocation line, leaving a hollow core. The
image forces needed to achieve zero normal stress on the cylinder
surface produce contributions to the classical Volterra displacement
field [14]. In real solids the problem is considerably more complex
as the core is not hollow but contains material which is stressed
into a highly nonlinear elastic state. However, the problem may be
approached in the same way as the hollow core dislocation, i.e., a
linear elastic contribution to the Volterra field is expected as a
result of tractions along the core boundary which accommodates the
nonlinear elastic behavior within the core. This core field must
modify the long-range linear-elastic interactions between dislocations and point defects predicted solely on the basis of Volterra
field.

The properties of the core field have been estimated by third-
order perturbation theory, first in a simple form by Zener [15]
and then by more rigorous and complex methods by Seeger and coworkers
[16-18]. This work leads to the prediction of a volume expansion
associated with the core field which is supported qualitatively by
measurements relating lattice parameter changes to dislocation
density in Cu [19]. However, because of the discrete nature of the
core and also because the strains are so large as to require even
higher order nonlinear elastic solutions, these analytical methods
are not adequate to completely display details of the core field.

An alternative is to model the core by atomic simulation of the
defect in regions containing discrete atoms interacting according to
a nonlinear force law. Here the problems of the continuum models

arising from nonlinear complications and the discreteness of the core
are entirely circumvented. Instead, the atomic simulation methods
are limited only by our incomplete understanding of the interatomic
force laws and by some assumptions made in the computational methods.
Of particular concern are the boundary conditions on the atomic
region, because for practical reasons the models contain only a few
thousand atoms. If this small discrete region is to behave as if
it were imbedded in a macroscopic crystal, the location of the
outermost atoms must be determined by the displacement field appropriate for an effectively infinite linear elastic solid. The bulk
of previous simulation work on dislocations treats the boundary
atoms as fixed in the positions predicted by the classical Volterra
field (a review of this subject can be found in Refs. [20] and [21]).
In this unrealistic procedure, the core field is cancelled at the
boundary. Recently, however, flexible boundary techniques [22,23]
have been added to the simulation calculations, making it possible
to examine the core field without the artificial constraints imposed
by fixed boundaries.

Extensive use of flexible boundary conditions has been made in
models of edge dislocations in α-Fe [24] and KCl [25] as well as for
cracks in α-Fe [26]. Much of this work has been specifically directed to the description of long-range contributions arising from
the core field which are important in solute interactions. Accordingly, in this paper we review the essential findings of this work
and discuss the implications to point defect-dislocation interactions. These results make possible a more accurate description of
the forces acting on point defects near the dislocation core than can
be made with the Volterra field, but remain inadequate to describe
the consequences of placing a defect within the core. Within the
core region atomic calculations of the type described by Perrin,
et al. [27] are necessary.

3. THE ATOMIC MODEL OF A DISLOCATION

The atomic simulation procedure has been described in detail
elsewhere [21,24,25,28] and only a brief summary is given. Our
review of the core field results is based on work presented in
Refs. [24] and [25] in which infinitely long, straight edge dislocations were modeled. Therefore, the resultant displacement fields
are two-dimensional. These results were derived from α-Fe models
in which the dislocation is of the type < 100 > {100} and from KCl
models with the < 110 > {110} dislocation. The Johnson potential
[30] was employed to model α-Fe while a coulombic-plus-core-overlap
repulsion interaction of the type described by Tosi [31] was employed
to represent KCl. The discrete atomic region of the model is typically centered on the dislocation origin and contains atoms which
are free to move during the relaxation process. This process

consists of simultaneously solving the equations of motion of these
interior atoms over a period of small time increments. The kinetic
energy which appears as these atoms approach their equilibrium positions is constantly extracted. At occasional intervals the boundary
atoms are adjusted to match the constantly changing displacement
field of the interior region to that of an elastic continuum to which
it is connected. Three flexible boundary techniques to perform this
adjustment have been developed [22,24,29]. The computations are
ended when there are zero net forces (actually forces of the order
of the round-off error in the computation) within both the interior
atomic and external-continuum regions of the model.

4. THE DISLOCATION FIELD ARISING FROM THE CORE

Results for the core field displacements are illustrated in
Fig. 1 which shows the central region of the KCl model containing a
< 110 > {110} edge dislocation. The ions are shown in positions
predicted by the isotropic Volterra field, and the vectors represent
the net displacement of each ion as the model achieves equilibrium.
The final results are independent of whether the ions are initially
positioned by the isotropic or anisotropic version of the Volterra
field although the isotropic form was found to be a somewhat closer
representation of the equilibrated state. A clearer description
of the core field is contained in Fig. 2 which is derived from the
α-Fe results and in Fig. 3, taken from the KCl results. Here the
cylindrical u_r and u_θ components of the field were evaluated for
atoms which were within ± 1 Å of the circle of radius indicated in
these figures.

Although it is not shown here, this displacement field varies
dramatically very close to the dislocation. However, outside the
nonlinear core region (1b to 2b away from the origin) the field was
found to assume approximately an r^{-1} dependence, although some contributions from terms containing r^{-n} where n > 1 are present, particularly in the KCl models. A consequence of the r^{-1} behavior is
that the core field produces a net dilatation of the lattice. The
magnitude of this dilatation is given by the expansion area per unit
length of the dislocation and is found by numerical integration of

$$\delta A = \int_0^{2\pi} r \, u_r \, d\theta . \tag{1}$$

Values of δA evaluated along the largest circles contained in these
two models are given in Table I. These models predict net expansions
of the order of 0.5 to 1.4 atomic volumes per repeat distance along
the dislocation line. These values are in reasonable agreement with
previous estimates [19].

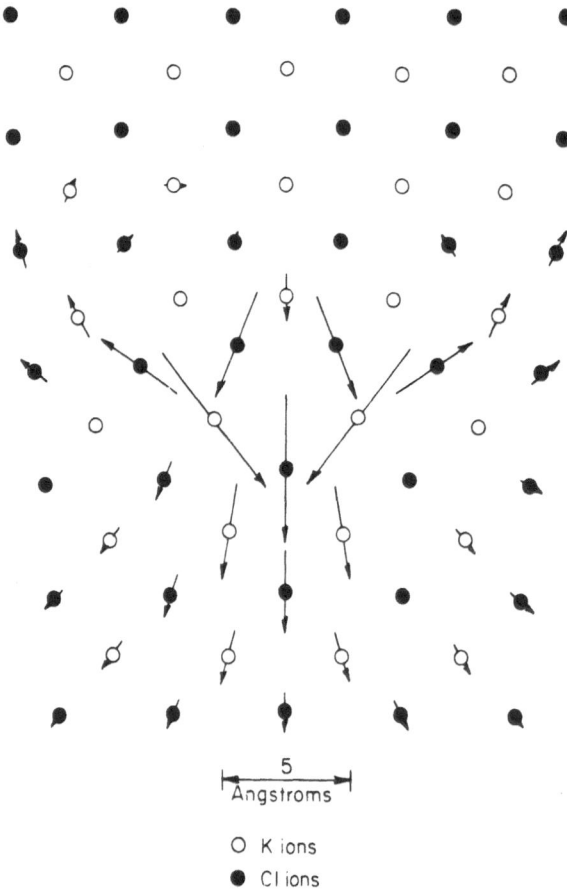

Fig. 1. Ion rearrangements in the central portion of a KCl model.
The ions are located according to the isotropic Volterra displacement field and the vectors represent the net displacement (magnified
10 times) of each ion as a result of relaxation to the equilibrium
arrangement. From Hoagland [25].

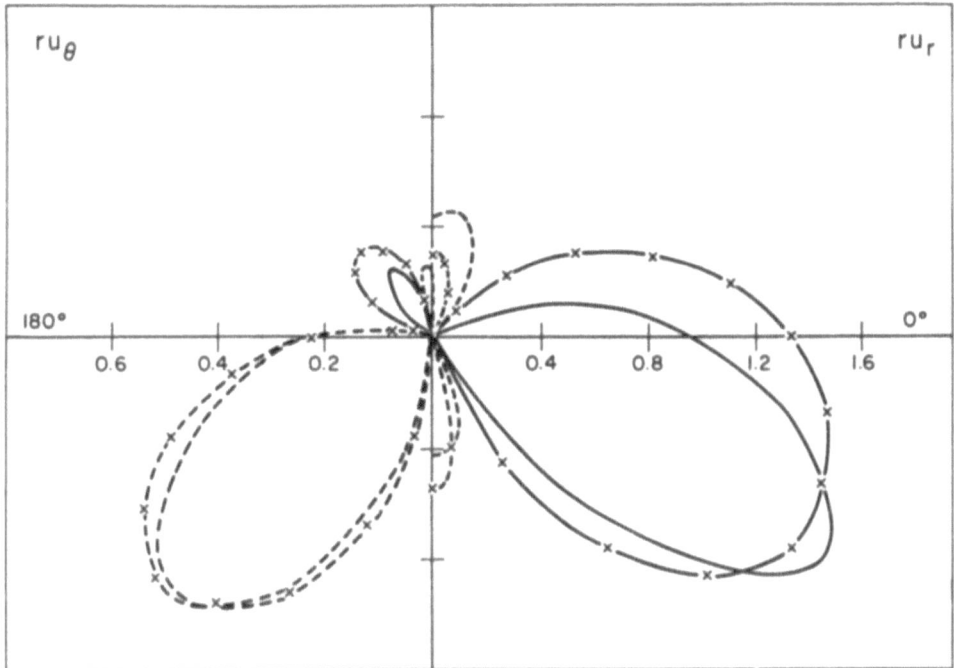

Fig. 2.   Polar plots of the ru_r (right side) and ru_θ components of
the core displacement field at an average distance of r = 17.5 Å
from the dislocation line in a model of α-Fe. Negative values are
shown dashed. Curves denoted by "x" are calculated from Eqn. (2)
using data given in Table I. From Gehlen, et al. [24].

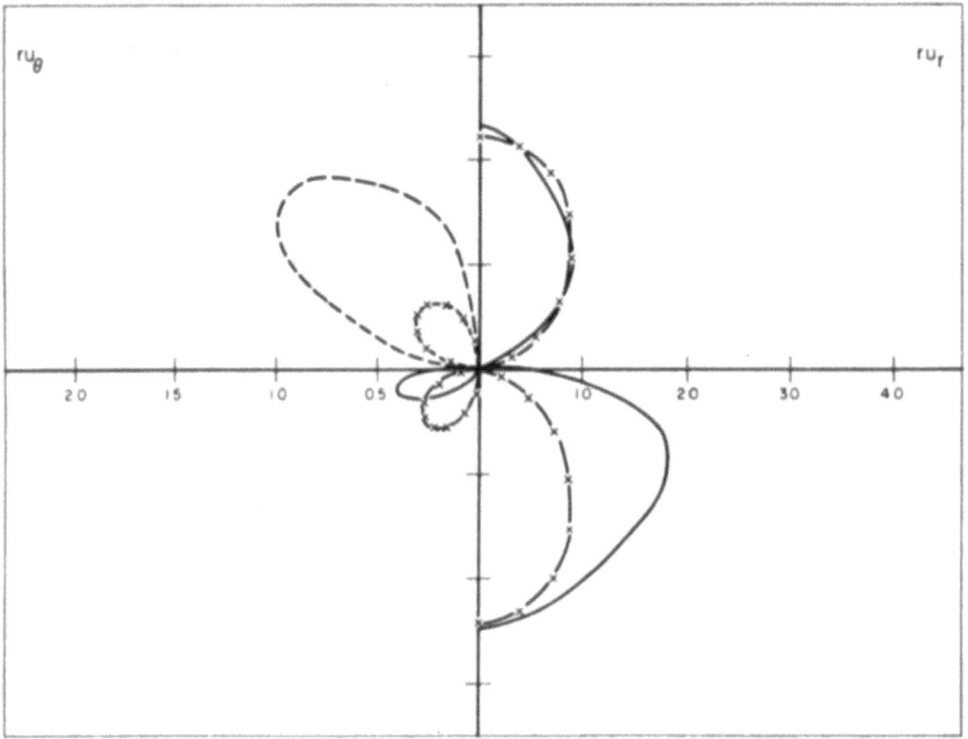

Fig. 3.   Polar plots of the ru_r (right side) and ru_θ components of
the core displacement field at an average distance of r = 24 Å in
a KCl model. Negative values are shown dashed. Curves denoted by
"x" are calculated from Eqn. (2) using data given in Table I.
From Hoagland [25].

Table I. Core Field Parameters

Model	Expansion Area (δA Computed by means of Eqn. (1)	Eccentricity [a] Ratio (α)	Distance of Dilatation Center Below Dislocation Line [a]
α-Fe [24]	$0.25b^2$	3.33	1.8b
KCl [25]	$0.5b^2$	0.5	0.3b to 3.2b

(a) Values which give the best fit of the core field to the dilatation field represented by Eqn. (2)

A useful approximation to the core field was found to be an elliptical center of expansion. The source of the expansion is composed to two orthogonal, unequal pairs of dipole forces. In the isotropic form, the displacements for this defect are

$$u_r = \frac{\delta A}{2\pi} \left(1 + \frac{\alpha - 1}{\alpha + 1} \frac{\cos 2\theta}{1 - \beta} \right) \quad ,$$

$$u_\theta = \frac{\delta A}{2\pi r} \left(\frac{1 - \alpha}{1 + \alpha} \right) \sin 2\theta \quad , \tag{2}$$

where $\beta = 1/[2(1 - \nu)]$ and α is the ratio of the strengths of the force dipole in the x-direction (parallel to glide plane) to the dipole along the y-direction. The angle θ is measured from the glide plane. The field represented by Eqn. (2) was found to be a very good approximation to the α-Fe core field when the dilatation source was placed approximately 5 Å below the dislocation origin (the extra half-plane of atoms is on the y > 0 axis). By suitably adjusting both the value of α and the separation between the dilatation source and the dislocation, a best fit shown in Fig. 2 was obtained. The same precision in the fit of Eqn. (2) to the KCl results could not be achieved although it gives a reasonable representation of the core field, as can be seen in Fig. 2. Here again, the results require that the expansion field be located beneath the dislocation origin. Values of α and the distance of separation are recorded in Table I for these two materials. The lack of precision in the fit of Eqn. (2) to the KCl results is reflected in the uncertainty in the separation distance.

Some additional consequences of the core field pertinent to point defect interaction were found by Hoagland [25]. Most importantly, the dilatation components of the core field give rise to a substantial decrease in the elastic constants near the dislocation.

This is revealed by comparison of the energy factor for a dislocation
in an infinite continuum and the results of the atomic calculations.
The energy factor, K, is the grouping of stiffness constants in the
coefficient of the ℓnR term in the well known expression for the
excess internal energy produced on introducing a dislocation in a
solid [32]. In the anisotropic case, the energy factor for the
Volterra field becomes

$$K = (C_{11} + C_{12}) \left(\frac{C_{66}(C_{11} - C_{12})}{C_{11}(C_{11} - C_{12} + 2C_{66})} \right)^{1/2} , \qquad (3)$$

which reduces to

$$K = \frac{\mu}{1 + \nu} ,$$

in the isotropic limit. From Eqn. (3) the value of K appropriate
for the interatomic potential in the KCl calculations is 0.089 eV/Å^3
(0.097 eV/Å^3 assuming isotropy). However, the results of the model
calculations, which are contained in Fig. 4, indicate that the measured K in the core region decreases from 0.117 eV/Å^3 to 0.069 eV/Å^3
as the model is allowed to relax from a state where the ions are
positioned by the Volterra field to the equilibrated condition.
These K values are based on the summation of the excess potential
energy of the lattice up to a distance of about 13b from the dislocation. Therefore, this behavior points to a significant local
softening accompanying the creation of the core field. While the
size of the "softened" region could not be ascertained, at sufficiently large distances from the dislocation, K must assume the value
appropriate for a linear elastic continuum.

5. DISLOCATION-SOLUTE ATOM INTERACTIONS

Although the details of the core field for the α-Fe and KCl
model are not identical, the results show some striking similarities.
In view of the differences in the interatomic potential and the crystallographic structure in these two models, we anticipate that these
results will be essentially applicable to crystalline materials in
general. Hence, consequences with respect to solute-dislocation
interactions are expected for all crystals.

The principal effect of the core field is to alter the size and
modulus interactions which determine the equilibrium point defect
concentration around dislocations. Eqn. (2) suffices to provide some
estimates of these effects in the case of edge dislocations. If
we also assume for simplicity that the dilatation center and the dislocation line are coincident, then the stresses of the dilatation
field become

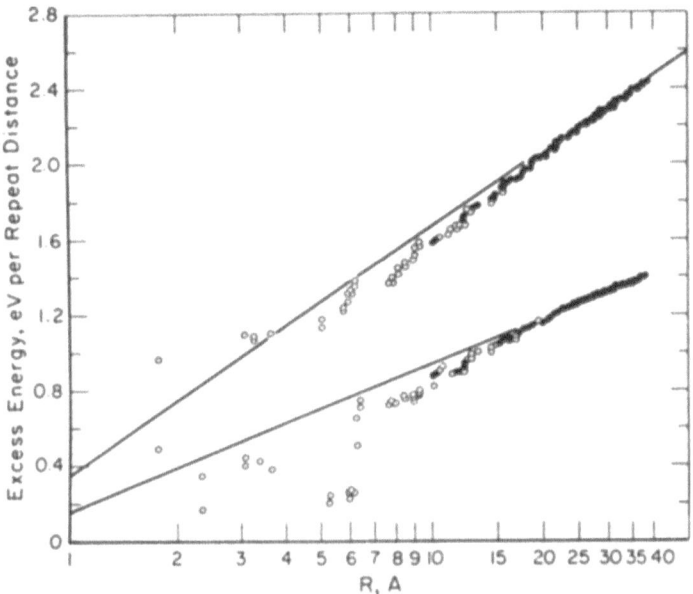

Fig. 4. Excess potential energy of the lattice contained within
cylinders of radius R centered on the dislocation origin in KCl.
The upper data are the unrelaxed Volterra array and the lower, the
equilibrated array results. The slopes of the lines drawn tangent
to the linear portions are equal to $Kb^2/4\pi$.

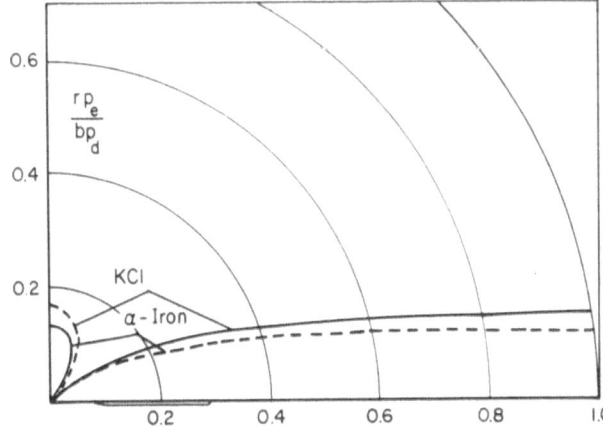

Fig. 5. A comparison of the relative magnitude of the hydrostatic
pressure component of the dilatation field, P_e, to that of the
dislocation field, P_d, in terms of the dimensionless ratio rP_e/bP_d.
This ratio, shown here in polar form in the first quadrant, was
derived from the KCl and α-Fe models. Negative values are indicated
by dashed line.

$$\sigma_{rr} = - \frac{\mu \delta A}{\pi r^2} \ [1 + \frac{2\beta(\alpha - 1)}{(1 + \alpha)(1 - \beta)} \ \cos 2\theta] \quad ,$$

$$\sigma_{\theta\theta} = \frac{\mu \delta A}{\pi r^2} \quad ,$$

$$\sigma_{r\theta} = \frac{\mu \delta A}{\pi r^2} \ \frac{(1 - \alpha)}{(1 + \alpha)(1 - \beta)} \ \sin 2\theta \quad ,$$

$$\sigma_{zz} = \nu(\sigma_{rr} + \sigma_{\theta\theta}) \quad . \tag{4}$$

During formation, the purely dilatational component of the strain field of a point defect does work against the hydrostatic pressure component of the dilatation field obtained from Eqn. (4) and given by

$$P_e = - \frac{\mu \delta A}{3\pi r^2} \ \frac{(4\beta - 1)(\alpha - 1)}{(1 - \beta)(1 + \alpha)} \ \cos 2\theta \quad , \tag{5}$$

and also against the pressure component P_d of the Volterra field. The net hydrostatic pressure is thus

$$P_e + P_d = \frac{\mu b(4\beta - 1)}{3\pi r} \ [\frac{\sin \theta}{2\beta - 1} - \frac{ab(\alpha - 1) \cos 2\theta}{r(1 - \beta)(1 + \alpha)}] \quad , \tag{6}$$

where the first term in the bracket is the dislocation contribution and we have set $\delta A = ab^2$. The relative magnitude of the hydrostatic pressure produced by the dilatation field may be compared to that of the dislocation field in terms of the dimensionless ratio rP_e/bP_d. This ratio is given in Fig. 5 as a polar plot in the first quadrant for the KCl and α-Fe results (this ratio is symmetric and antisymmetric with respect to the y and x axes, respectively). As would be expected, the dilatation field dominates the hydrostatic pressure component of the combined dilatation-dislocation field near the glide plane. We see also that for $\alpha < 1$, the KCl result, the dilatation field tends to counteract the Volterra field immediately above the origin but enhances this field below it. The opposite effect is true for Fe where $\alpha > 1$.

In dilute solutions the distribution of defects in equilibrium with the Volterra field would be modified by the superposition of the dilatation field, and from Eqn. (6) we obtain

$$C = C_d \exp \left[- \frac{D_e (\alpha - 1) \cos 2\theta}{r^2 kT} \right] \quad , \tag{7}$$

where

$$D_e = \frac{\mu ab^2 (4\beta - 1) \delta V}{3\pi (1 - \beta)(1 + \alpha)} \quad ,$$

C_d = the concentration of defects in equilibrium with the Volterra field given by

$$C_d = C_o \exp \left[- \frac{D_d \sin \theta}{rkT} \right] \quad , \quad \text{and}$$

δV = the local volume change at the defect site. Since $0.5 < \beta < 1$, D_e has the same sign as δV and therefore the relative effect on the local defect concentration is determined by the sign of the term $(\alpha - 1)$.

In the case of vacancies (or interstitials), symmetry requires that the dilatation field cannot experience an osmotic force [33]. However, the diffusion-controlled climb rate which is proportional to the local concentration of vacancies [33] can be influenced by the presence of the dilatation field for $\alpha \neq 1$. The following derivation serves to illustrate this effect. Again considering only the dilatational component of the defect, the total capacity of the dilatation field for defects above that which would be in equilibrium with the Volterra field can be obtained by expanding Eqn. (7). If $D_d(\alpha - 1) < r_o^2 kT$ and $D_e < r_o kT$, then to second order we obtain

$$C - C_o = - \frac{D_e C_o \sin \theta}{rkT} - \frac{D_d(\alpha - 1)C_o \cos 2\theta}{r^2 kT} + \frac{D_e^2 C_o \sin^2 \theta}{r^2 k^2 T^2}$$

$$+ \frac{D_d^2 (\alpha - 1)^2 C_o \cos^2 2\theta}{2r^2 k^2 T^2} + \frac{D_e D_d (\alpha - 1)C_o \cos 2\theta \sin\theta}{r^3 k^2 T^2}. \tag{8}$$

The net number of extra defects per unit length of dislocations is thus

$$\frac{N}{L} = \int_{r_o}^{R} r dr \int_0^{2\pi} (C - C_o) \, d\theta \quad .$$

The cross term and the first two terms on the right side of Eqn. (8) integrate to zero leaving

$$\frac{N}{L} = \frac{N_e}{L} + \frac{N_d}{L} \cong \frac{\pi D_e^2 C_o}{2k^2 T^2} \ln \frac{R}{r_o} + \frac{\pi D_d^2 (\alpha - 1)^2 C_o}{2 r_o^2 k^2 T^2} \quad .$$

Near the dislocation where $\ln R/r_o$ is of the order of unity, the increased capacity for defects due to the dilatation center is reflected by the ratio

$$\frac{N_e}{N_d} \cong \frac{D_d^2 (\alpha - 1)^2}{D_e^2 r_o^2} = a^2 \left(\frac{b}{r_o} \right)^2 \left(\frac{2\beta - 1}{1 - \beta} \right)^2 \left(\frac{\alpha - 1}{\alpha + 1} \right)^2 \quad . \tag{9}$$

Using the data in Table I (and taking $r_o \sim b$, and $\beta = 2/3$) we find for both the α-Fe and KCl models that the net increase in defects in this region is small, amounting to only about 3%. On the other hand, we can expect that for dislocations in materials which possess a large δA, the dilatation may contribute to a local increase in defects which is comparable to that afforded by the Volterra field. Similar analyses along the lines described by Fleischer [34] can be derived which include the interaction between the core field and the asymmetric component of the lattice distortion at the defect site.

A larger effect is produced directly by the "softening" behavior of the elastic constants in the core region. To first approximation the interaction energy is proportional to the difference in elastic constants between the defect region and the bulk solvent and in the simple case of interaction with the hydrostatic component of internal stress, Hirth and Lothe [35] give

$$\Delta W_m = P^2 V_a \frac{B - B'}{BB'} \quad ,$$

where V_a is the effective volume of the defect and B and B' are the bulk moduli of the defect-free and defect regions of the lattice, respectively.

On the basis of the simulation results, the strength of the modulus interaction should be modified as the defect approaches a dislocation because the region in the vicinity of the dislocation is elastically softer than its surroundings. In fact, from the KCl results we see that the energy factor, K, is approximately 25% less than expected in a cylinder radius of 13b containing the dislocation. This points to a very significant reduction in the local elastic constants, and, therefore, whether the modulus interaction is attractive or repulsive, the effect of local softening is to diminish the interaction strength. In fact, this 25% decrease in the elastic constants of the near-core region would produce a similar change in local solute concentration according to Eqn. (8).

Finally, other interactions deriving from the presence of the core field may also be important. For example, in metals the decrease in Fermi energy accompanying the expansion could give rise to much longer range electronic interactions than predicted solely from the Volterra field. In this regard, to maintain a constant electron chemical potential throughout the solid, the Volterra displacement field leads to a redistribution of conduction electrons creating a dipole [11], while the core field should generally increase the number of conduction electrons near the dislocation [5]. In addition, since the core field arises from nonlinear effects, the presence of a defect in the core could alter this nonlinear behavior. Substantial modifications of the long-range field arising from the core at these locations along the dislocation line would result. While this is a rather difficult problem to treat, it is clear that it can be studied through atomic calculations of the type described here.

In summary, the isotropic treatment of the long-range field arising from core nonlinearities shows that effects are produced completely parallel to those associated with the Volterra field. Evidently, a similar parallelism exists in the anisotropic elastic case: for example, the nonlinear field could then include shear stress terms producing added interactions with a tetragonal defect such as C in ferrite. Within the next few years burgeoning information on interactions arising from nonlinearities is expected.

ACKNOWLEDGMENTS

This work was supported by the Air Force Office of Scientific Research under Grant AFOSR-71-2082.

REFERENCES

1. ESHELBY, J.D., Solid State Phys., 3 79 (1956).
2. SUZUKI, H., Sci. Repts. Tohoku Univ., A4 455 (1952).
3. BESHERS, D.N., Acta Met. 6 521 (1958).
4. HIRTH, J.P. and LOTHE, J., Theory of Dislocations, McGraw-Hill, New York, 1968.
5. NABARRO, F.R.N., Theory of Crystal Dislocations, University Press, Oxford, 1967.
6. FRIEDEL, J., Dislocations, Addison-Wesley, New York, 1967.
7. MASUMURA, R.A. and SINES, G., J. Appl. Phys. 41 3930 (1970).
8. BARNETT, D.M., phys. stat. sol. 496 741 (1972).

9.   BARNETT, D.M. and SWANGER, L.A., phys. stat. sol. 48b 419 (1971).
10. BARNETT, D.M., LOTHE, J., NISHIOKA, K. and ASARO, R.J., J. Phys. F: Metal Phys. 3 1083 (1973).
11. COTTRELL, A.H., HUNTER, S.C. and NABARRO, F.R.N., Phil. Mag. 44 1064 (1953).
12. TILLER, W.A. and SCHRIEFFER, R., Scripta Met. 4 57 (1970).
13. HIRTH, J.P. and COHEN, M., Met. Trans. 1 3 (1970).
14. LOVE, A.E.H., The Mathematical Theory of Elasticity, Cambridge Univ. Press, Cambridge, 1927, p. 227.
15. ZENER, C., Trans. AIME 147 361 (1941).
16. SEEGER, A. and MANN, E., Z. Naturforschung 14a 154 (1959).
17. TEODOSIU, C., Fundamental Aspects of Dislocation Theory, NBS Special Publ. 317, 389 (1970).
18. TEODOSIU, C. and SEEGER, A., Fundamental Aspects of Dislocation Theory, NBS Special Publ. 317, 877 (1970).
19. CRUSSARD, C. and AUBERTIN, F., Rev. Met. (Paris) 46 354 (1949).
20. Interatomic Potentials and Simulation of Lattice Defects, GEHLEN, P.C., BEELER, J.R. and JAFFEE, R.I., eds., Plenum Press, New York, 1972.
21. BEELER, J.R., Advances in Materials Research, 4, HERMAN, H., ed., Interscience, New York, 1970, p 295.
22. SINCLAIR, J.E., J. Appl. Phys. 42 5321 (1971).
23. GEHLEN, P.C., in Interatomic Potentials and Simulation of Lattice Defects, GEHLEN, P.C., BEELER, J.R. and JAFFEE, R.I., eds., Plenum Press, New York, 1972, p 475.
24. GEHLEN, P.C., HIRTH, J.P., HOAGLAND, R.G. and KANNINEN, M.F., J. Appl. Phys. 43 3921 (1972).
25. HOAGLAND, R.G., Ph.D. Thesis, The Ohio State University, 1973.
26. GEHLEN, P.C., Scripta Met. 7 1115 (1973).
27. PERRIN, R.C., ENGLERT, A. and BULLOUGH, R., in Interatomic Potentials and Simulation of Lattice Defects, GEHLEN, P.C., BEELER, J.R. and JAFFEE, R.I., eds., Plenum Press, New York, 1972, p 509.
28. GEHLEN, P.C., ROSENFIELD, A.R. and HAHN, G.T., J. Appl. Phys. 39 5246 (1968).
29. GEHLEN, P.C., HIRTH, J.P. and HOAGLAND, R.G., to be published.
30. JOHNSON, R.A., Phys. Rev. 145 423 (1966).
31. TOSI, M.P., Solid State Phys. 16 1 (1964).
32. HIRTH, J.P. and LOTHE, J., in Theory of Dislocations, McGraw-Hill, New York, 1968, p 417.
33. HIRTH, J.P. and LOTHE, J., in Theory of Dislocations, McGraw-Hill, New York, 1968, p 510.
34. FLEISCHER, R.L., in Strength of Metals, PECKNER, D., ed., Reinhold, New York, 1965, p 93.
35. HIRTH, J.P. and LOTHE, J., in Theory of Dislocations, McGraw-Hill, New York, 1968, p 469.

COMPUTER EXPERIMENTS ON IMPURITY EFFECTS IN ALPHA-IRON

J. R. Beeler, Jr.

Nuclear Engineering Department
North Carolina State University
Raleigh, North Carolina   27607

ABSTRACT

A series of dynamical computer experiments was performed on mechanical property changes produced by complexes consisting of impurity atoms in bound states with other lattice defects. Complexes based on interstitial and substitutional impurities were studied. Major emphasis was placed on the response of irradiated samples. Hence, complexes involving self-interstitials are treated in more detail than is the case for radiation-free environments. Strain tensor components are given for tetragonal complexes and building modules for ordered defect structures are described.

1.   INTRODUCTION

This paper concerns some particular computer experiment results on the effects of substitutional and interstitial impurities in α-Fe. The substitutional impurities treated are He and two 'size-effect' metallic impurities. The size-effect metallic impurities are 10% oversize and 10% undersize metallic atoms. The interstitial impurity is C. Johnson's interatomic potentials were used in the case of C [1], Fe [2], and the two metallic impurities [2]. The interatomic potential of Wilson and Bisson [3] was used in the case of He.

Two main points of interest dominate the presentation. One point concerns dislocation pinning by complexes comprised of a few impurity atoms and a few vacancies or self-interstitials. The other point concerns the role of these complexes as nucleation centers and/or building modules for ordered defect structures. The computer experiment programs used were designed to provide the

117

atomic displacement field associated with a given complex and the
configuration, binding and migration energies of this complex.
Knowledge of the displacement field for a given complex allows one
to set up a strain tensor and principal coordinate system which then
can be used, in the type of formalism developed by Cochardt et al.
[4] for C in α-Fe, to estimate dislocation pinning strength for the
complex. Strain tensors are given for those complexes with binding
energies sufficiently large to merit consideration at a finite temperature. A direct statement of the displacement field is made for
the primary point defects.

The principal motivation for this work was to develop a com-
pilation of the properties of vacancy-impurity and self-interstitial-
impurity complexes for use in engineering calculation on radiation
effects in fission and fusion reactor radiation fields. In these
two instances, the self-interstitial is an important defect in
routine engineering calculations, whereas this defect is seldom of
practical importance in conventional (radiation-free) environments.

Strengthening (hardening) processes in nuclear reactor component materials have all of the traditional aspects in common with
strengthening processes in materials in radiation-free environments.
In addition, however, there are three novel aspects which are pe-
culiar to strengthening processes in a radiation environment. One
of these is the continual production of vacancies and interstitials
at rates which are enormously larger than those that can be induced
solely by thermal excitation. The second is the production of types
of defects which are not produced by thermal excitations. A third
novel aspect is the resolution of impurity atom precipitates by
atomic collision cascades at temperatures less than the normal dissociation temperature.

Experiments show that certain substitutional impurity atoms
make some steels more susceptible to brittle fracture in a radiation
environment than in a radiation-free environment. However, the
mechanisms for this embrittlement are neither formulated nor understood. Hence, one prime motivation for our computer experiment
studies is to characterize mechanisms for substitutional impurity
effects on solid solution strengthening and on alterations of dispersion strengthening in a radiation environment.

2. COMPUTATIONAL METHOD

Impurity atom effects in α-Fe were studied using the dynamical
computer experiment method introduced by Gibson et al. [5] In this
method, atom movements in a finite crystal containing N atoms are
computed directly by numerically integrating the 3N equations of
motion involved. Within the limits of classical mechanics applied
to atom motions in a crystal or fluid, this method gives a complete

many-body description of mechanical behavior within the crystal
studied. Static equilibrium atom configurations associated with a
given lattice defect can be obtained by using artificial damping
techniques [6].

In the present work, the interatomic forces for Fe-Fe atom interactions were obtained from Johnson's [2] potential for α-Fe.
This potential is his central force potential for Fe. A Johnson
potential also was used to describe the interaction between a C
and an Fe atom [1]. The Wilson-Bisson potential was used to describe the interaction between a He and an Fe atom [3].

A 10% misfit metallic small impurity atom (SIA) was simulated
by shifting Johnson's potential for α-Fe toward smaller interatomic
distances by the appropriate amount. Similarly, a 10% misfit large
impurity atom (LIA) was simulated by shifting Johnson's potential
for α-Fe toward larger interatomic distances by the appropriate
amount. These misfit impurities are 'metallic' in character. There
is a slight modulus effect in this model but the main influence is
that of a 'size-effect' metallic impurity in each case.

The He-Fe interaction described by the Wilson-Bisson potential
is wholly repulsive. The C-Fe interaction described by Johnson's
potential for Fe-C is a strong, short-range chemical interaction of
a highly covalent character. Its strength is 0.35 eV and the bond
length is 1.94 Angstroms. This potential was constructed to give
the correct geometry and binding for Fe_3C (cementite) in α-Fe.

3. MISFIT METALLIC IMPURITIES

Impurity atom effects on mechanical properties are intimately
connected with the properties of vacancies and self-interstitials,
the primary point defects. This is true because impurity atoms tend
to form complexes with the primary point defects.

From the standpoint of strengthening mechanisms, the most
important property of a point defect is its displacement field.
The small impurity atom (SIA) and the large impurity atom (LIA)
produce displacement fields which are listed in Tables I and II,
respectively. Note that all displacements for the SIA are inward,
toward the impurity, and that all displacements for the LIA are
outward. The displacement field for a vacancy (V) is listed in
Table III. In this instance, successive neighbor shells are displaced alternately inward and outward. This alternating displacement sense of the vacancy displacement field allows strain energy
relief for both a SIA and a LIA when either of them is situated near
a vacancy. In all three instances, the 8-th and 9-th neighbor
shells show no displacement, and those of the 24 10-th neighbors
of the vacancy, that are not situated on a closepacked line, show

Table I. Displacement Field for a SIA (small impurity atom) in
α-Fe. n is the neighbor shell index. AN is the atom number in the
DYNAM computational shell, and $\vec{r}_p$ is the associated site position
vector relative to the SIA position at the origin. Distances and
displacements are expressed in hlc (half lattice constants).

n	AN	$\vec{r}_p$	Δx	Δy	Δz	Δr	Sense
* 1	1282	[111]	−.015	−.015	−.015	0.027	in
2	1167	[200]	−.003	0	0	0.003	in
3	1178	[220]	−.004	−.004	0	0.005	in
4	1283	[311]	−.002	−.001	−.001	0.003	in
* 5	1399	[222]	−.006	−.006	−.006	0.010	in
6	1168	[400]	0	0	0	0	--
7	1293	[331]	−.002	−.002	−.001	0.003	in
8	1180	[420]	0	0	0	0	--
9	1401	[422]	0	0	0	0	--
* 10	1514	[333]	−.002	−.002	−.002	0.004	in
10	1284	[511]	0	0	0	0	--
*4d_1	1632	[444]	−.001	−.001	−.001	0.002	in
*5d_1	1746	[555]	−.0	−.0	−.0	0.001	in
*6d_1	1865	[666]	−.0	−.0	−.0	0.000	in

*Close-packed line direction -- SIA at Site 1166.   Run NC-1356

Table II. Displacement Field for a LIA (large impurity atom) in
α-Fe. n is the neighbor shell index. Atom numbers (AN) and site
position vectors ($\vec{r}_p$) are given in Table I. Distances and displacements are expressed in hlc (half lattice constants).

n	Δx	Δy	Δz	Δr	Sense
* 1	0.019	0.019	0.019	0.033	out
2	0.008	0	0	0.008	out
3	0.005	0.005	0	0.007	out
4	0.003	0.002	0.002	0.005	out
* 5	0.007	0.007	0.007	0.013	out
6	0.001	0	0	0.001	out
7	0.003	0.003	0.001	0.004	out
8	0	0	0	0	--
9	0	0	0	0	--
* 10	0.003	0.003	0.003	0.006	out
10	0.001	0	0	0.001	out
*4d_1	0.002	0.002	0.002	0.003	out
*5d_1	0.001	0.001	0.001	0.001	out
*6d_1	+0	+0	+0	0.001	out

*Close-packed line direction -- LIA at Site 1166. Run NC-1348

Table III. Displacement Field for a Vacancy in α-Fe. n is the
neighbor shell index. Atom numbers (AN) and site position vectors
($\vec{r}_p$) are given in Table I. Distances and displacements are expressed in hlc (half lattice constants).

n	Δx	Δy	Δz	Δr	Sense
* 1	$-.029$	$-.029$	$-.029$	0.051	in
2	0.051	0	0	0.051	out
3	$-.006$	$-.006$	0	0.009	in
4	0.004	0.003	0.003	0.005	out
* 5	$-.010$	$-.010$	$-.010$	0.017	in
6	0.007	0	0	0.007	out
7	$-.002$	$-.002$	0	0.003	in
8	0	0	0	0	--
9	0	0	0	0	--
* 10	$-.003$	$-.003$	$-.003$	0.006	in
10	0.001	0.001	0.001	0.002	out
*$4d_1$	$-.001$	$-.001$	$-.001$	0.002	in
*$5d_1$	-0	-0	-0	0.001	in
*$6d_1$	-0	-0	-0	-0	in

*Close-packed line direction -- Vacancy at Site 1166. Run NC-1355 negligible relaxation. Along close-packed lines, displacements greater than 0.001 hlc (half lattice constant) occur out to six interatomic distances $(6d_1)$. Here d_1 represents the nearest neighbor distance (first neighbor distance).

Table IV lists the basic vacancy-impurity atom complexes and their binding energies. A positive binding energy designates a bound system, whereas a negative binding energy designates a system which is not bound. Also given are the binding energies for the first six vacancy pair configurations, $V_2(n)$, for n = 1, 2, 3, 4, 5, 6. Here n is the neighbor shell index. In his pioneering work on defect property calculations, Johnson [7,8] showed that it is possible to estimate the binding energy of a vacancy cluster by adding the binding energies for all the vacancy-pair configurations in the cluster. In the present paper, it is shown that, although Johnson's additivity rule also applies to certain particular vacancy-impurity complexes, it does not apply to all types of vacancy-impurity complexes. This circumstance makes prediction of vacancy-impurity atom complex behavior more difficult than the prediction of vacancy cluster behavior.

Vacancy-SIA complexes are illustrated by Fig. 1(a) - 1(f) and two hybrid split-interstitial configurations by Fig. 1(g) and 1(h). A hybrid split-interstitial is one consisting of an impurity atom and a host metal atom situated as a pair about a normal atom position in a split-interstitial configuration. The computed binding energy (BE) is stated for each complex, in eV. The number in parentheses, below the computed binding energy statement, is the binding energy predicted by the additivity rule from the data in Table IV. The additivity rule worked well (to within 10%) in cases 1(b), 1(e), and 1(f), but did not work satisfactorily in cases 1(c) and 1(d). In those instances for which the additivity rule was good, vacancy-vacancy interactions were either weak or absent. In cases 1(c) and 1(d), for which the rule did not work, the binding energy is dominated by multiple, negative binding energy contributions from third and sixth neighbor vacancy pairs.

If one assumes that the negative binding interactions among vacancies are not changed by the presence of a SIA, then the existence of these multiple vacancy interactions decreases the effective V-SIA pair binding energy from 0.71 eV to 0.058 eV in case 1(c), and from 0.071 eV to 0.03 eV in case 1(d).

A complete treatment of impurity atom effects in an irradiated material must consider complexes made up of self-interstitials and impurity atoms and of split-interstitial configurations which include an impurity atom in their core. Any substitutional impurity atom displaced in an atomic collision cascade will terminate as a component of an interstitial configuration where it will remain until a vacancy or some other annihilatory sink is encountered. Table V

Table IV. Binding Energies for Selected Vacancy-Impurity
Complexes and Vacancy Pair (V_2) Configurations, α-Fe

Type of Complex	Shell Index	Binding Energy, eV	Type of Defect	Shell Index	Binding Energy, eV
V–SIA	1	−0.060	V_2	1	0.131
V–SIA	2	0.071	V_2	2	0.195
SIA–SIA	1	0.037	V_2	3	−0.03
SIA–SIA	2	0.058	V_2	4	0.04
V–LIA	1	0.082	V_2	5	−.01
V–LIA	2	−0.082	V_2	6	−0.03
LIA–LIA	all	neg.			

The SIA configuration energy is 0.099 eV.

The LIA configuration energy is 0.203 eV.

Table V. Configuration Energies for SIA and LIA
Hybrid Split Interstitials in α-Fe

Interstitial Orientation	Configuration Energy, eV	
	SIA	LIA
[110]	2.09	4.52
[111]	2.46	4.45

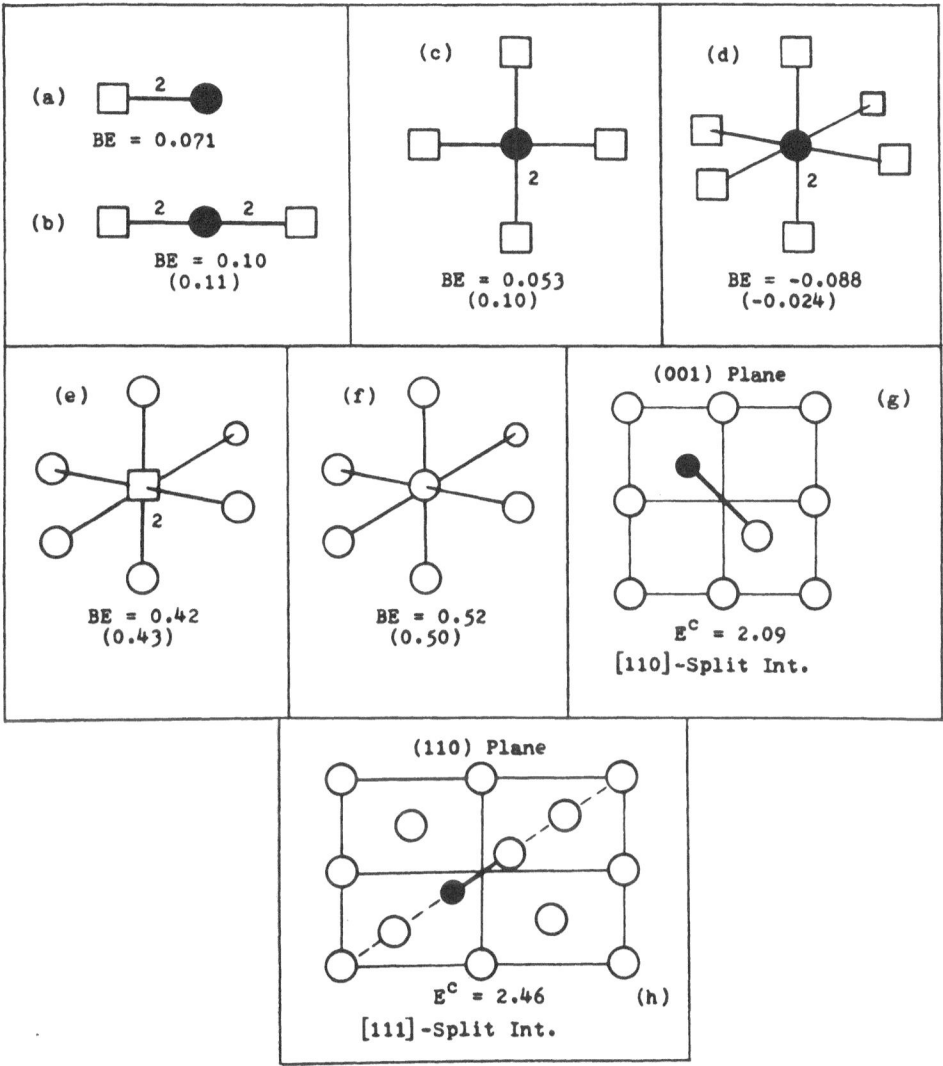

Fig. 1. Vacancy-small-impurity-atom complexes in α-Fe. Binding energy (BE) and configuration energy (E^C) are expressed in eV.

lists the hybrid split-interstitial configurations which are possible in α-Fe and their configuration energies. The lowest energy hybrid configuration for the SIA is a [110]-split interstitial with a configuration energy about 1 eV less than the [110]-split self-interstitial in α-Fe. The lowest energy hybrid configuration for the LIA is a [111]-split interstitial whose configuration energy is about 1.4 eV larger than the self-interstitial in α-Fe.

Segregation of SIA occurred in our computer experiments, but the size of a SIA cluster was limited to a six-atom octahedron shown in Fig. 1(f). All larger octahedron-shaped aggregates had negative binding energies.

On the basis of the magnitude of their configuration energies and binding energies, the most frequently occurring tetragonal defect complexes which involve a SIA should be configurations 1(a), 1(b) and 1(g). This statement applies to an irradiated material. The strain tensors for SIA configurations 1(a), 1(b), 1(f) and 1(g) are given in Table VI, along with the Cartesian coordinate system, u_i, for which they apply. The strain elements are given directly by the computer experiment program and follow the formalism of Cochardt et al. [4]

Vacancy-LIA complexes are illustrated by Fig. 2(a) - 2(g). The additivity rule worked only for cases 2(c) and 2(g). In all other instances, the binding energy predicted by the additivity rule exceeded the computed value by more than 10%. Neither pairs nor any larger aggregates of LIA exhibited positive binding energies.

On the basis of the magnitudes of their configuration energies and binding energies, the most frequently occurring tetragonal defect complexes which involve a LIA should be cases 2(a) and 2(c). As mentioned previously, the lowest energy hybrid split-interstitial involving a LIA is the [111] configuration. Strain tensors for configurations 2(a), 2(b) and the LIA split interstitial are given in Table VII.

Ordered defect structures of indefinite extent are possible in the case of the LIA. Platelet arrays made up of vacancies and LIA can extend indefinitely with a (100) plane orientation. These platelets (see Fig. 3) consist of a (001) plane of vacancies sandwiched between two adjacent (001) planes of LIA and Fe atoms. The binding energy per LIA is 0.40 eV. This structure is suggested by three circumstances: (1) the vacancy pair configuration with the largest binding energy is $V_2(2)$; (2) the vacancy-LIA pair with the largest binding energy is a first-neighbor pair; and (3) LIA pairs of either first or second-neighbor types have negative binding energies, whereas third-neighbor LIA pairs exhibit only a weak interaction.

Table VI. Strain Tensor Components for Tetragonal SIA Defects in α-Fe. Defect designationa 1a, 1b, 1f, and 1h refer to the parts of Fig. 1 which define SIA defect geometries. ε_i is the i-th strain component and u_i is the associated unit vector component for the strain coordinate system. $\vec{a}_d$ is the defect axis.

Defect	i	ε_i	u_i	Defect	i	ε_i	u_i
1a, 1b	1	0.051	$[100]$	1f	1	0.071	$[001]$
	2	-0.062	$[011]/2^{\frac{1}{2}}$		2	0.058	$[1\bar{1}0]/2^{\frac{1}{2}}$
	3	-0.062	$[0\bar{1}1]/2^{\frac{1}{2}}$		3	0.058	$[110]/2^{\frac{1}{2}}$
	$\vec{a}_d = [110]/2^{\frac{1}{2}}$				$\vec{a}_d = [001]$		

Defect	i	ε_i	u_i	Defect	i	ε_i	u_i
1h Hybrid Split Int.	1	0.292	$[001]$	Normal Split Int.	1	0.285	$[001]$
	2	-0.064	$[1\bar{1}0]/2^{\frac{1}{2}}$		2	-0.048	$[1\bar{1}0]/2^{\frac{1}{2}}$
	3	0.141	$[110]/2^{\frac{1}{2}}$		3	0.168	$[110]/2^{\frac{1}{2}}$
	$\vec{a}_d = [110]/2^{\frac{1}{2}}$				$\vec{a}_d = [1\bar{1}0]/2^{\frac{1}{2}}$		

Table VII. Strain Tensor Components for Tetragonal LIA Defects in α-Fe. Defect designations 2a and 2b refer to the parts of Fig. 2 which define LIA defect geometries. ε_i is the i-th strain component and u_i is the associated unit vector component for the strain coordinate system. $\vec{a}_d$ is the defect axis.

Defect	i	ε_i	u_i	Defect	i	ε_i	u_i
2a, 2b	1	-0.051	$[111]/3^{\frac{1}{2}}$	Hybrid Split Int.	1	0.60	$[111]/3^{\frac{1}{2}}$
	2	0.065	$[\bar{1}\bar{1}2]/6^{\frac{1}{2}}$		2	0.07	$[\bar{1}\bar{1}2]/6^{\frac{1}{2}}$
	3	0.042	$[1\bar{1}0]/2^{\frac{1}{2}}$		3	0.02	$[1\bar{1}0]/2^{\frac{1}{2}}$
	$\vec{a}_d = [111]/3^{\frac{1}{2}}$				$\vec{a}_d = [111]/3^{\frac{1}{2}}$		

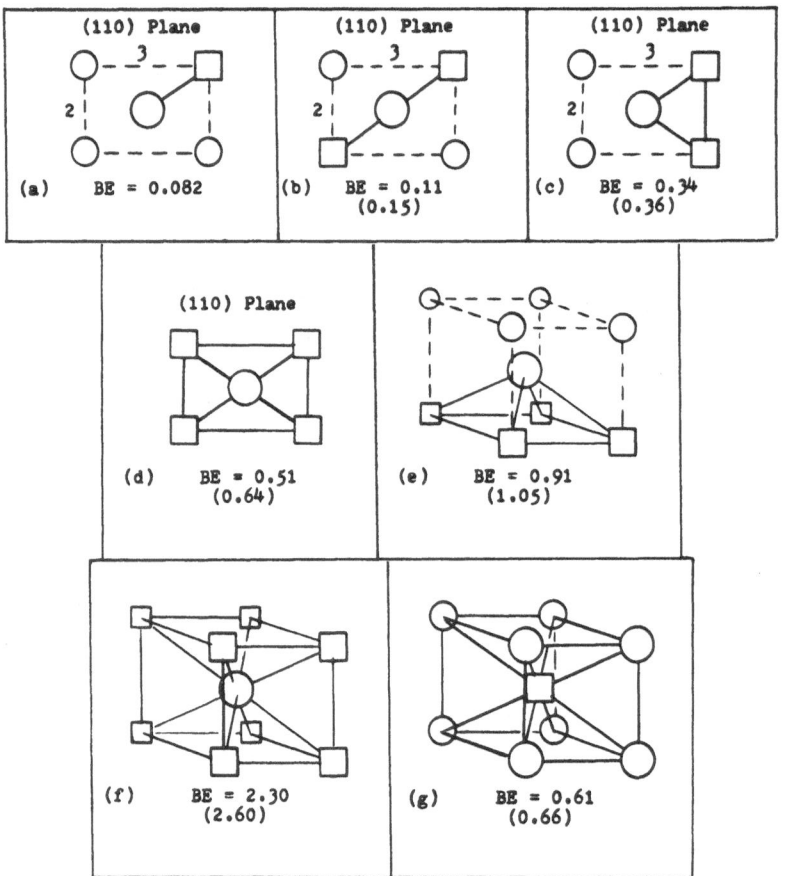

Fig. 2.   Vacancy-large-impurity-atom complexes in α-Fe.   Binding
energy (BE) and configuration energy (E^c) are expressed in eV.

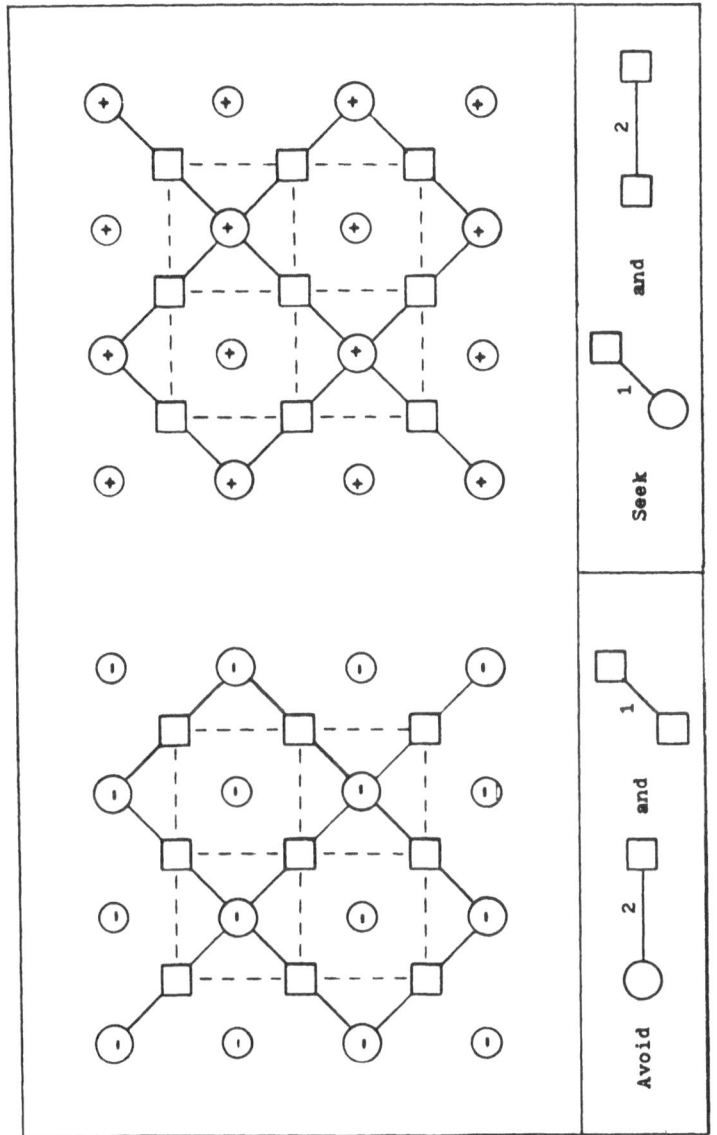

Fig. 3. Ordered defect structure of vacancies and large impurity atoms in α-Fe.

⊖ indicates atoms in plane z = -1 and ⊕ indicates atoms in plane z = 1. The vacancies are in plane z = 0.

Fig. 4 gives the configuration energy for V-SIA(n) and V-LIA(n) pairs in α-Fe. Fig. 5 gives the activation energy required to change from one configuration to another for neighbor indices n = 1 to 5. From these data, it is possible to construct the migration path geometry of a diffusing impurity atom at temperatures sufficiently low that the binding energy between a vacancy and an impurity atom could maintain correlated movements between a vacancy and an impurity atom.

Figs. 6 and 7 illustrate the correlated jump sequences between a vacancy and a SIA and LIA, respectively. In each instance, the correlated jump sequence can be confined to a (110) plane. In the case of the SIA, the vacancy must execute a five-jump sequence between successive movements of the impurity atom. The largest activation energy required is 0.68 eV, in the last step, which is also the activation energy for single vacancy migration in the specific computational model for Fe used in this paper. This circumstance means that the last step in the jump sequence is a uniform random event.

In the case of the LIA, the vacancy must execute a three-jump sequence between successive movements of the impurity atom. The largest activation energy required is 0.63 eV, the first step, which is less than the activation energy for vacancy migration in α-Fe. This circumstance makes a correlated movement sequence for a vacancy and a LIA possible at temperatures T such that kT is less than the binding energy between a vacancy and a LIA (0.082 eV).

We conclude that LIA can form extended defect structures with vacancies and that correlated movement of vacancies and LIA can take place. In contrast, SIA can segregate into tiny clusters of up to six SIA and must diffuse largely on a random vacancy access basis.

4. HELIUM IN ALPHA-IRON

Displacement fields and configuration energies for substitutional He, interstitial He and He-vacancy complexes in α-Fe were computed. This was done to study stabilization of void nuclei by He atoms. The interaction of octahedral He and of substitutional He with a void facet was studied in order to describe He migration into a void and He trapping at vacancies near a void facet. The simulations indicated that a He atom can be trapped by a vacancy and retained at any site near a (110) void facet excepting a site in the void facet itself. The configuration energy results for these cases are summarized in Table VIII.

In order to situate a He atom at a normal atom site, it is necessary, first, to produce a vacant site. In this context,

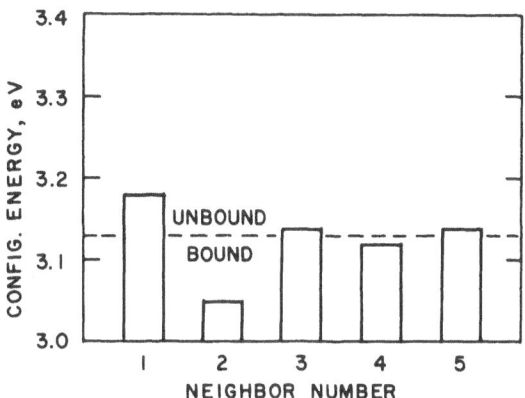

Fig. 4(a). Configuration energy for the Vacancy-SIA complex in α-Fe.

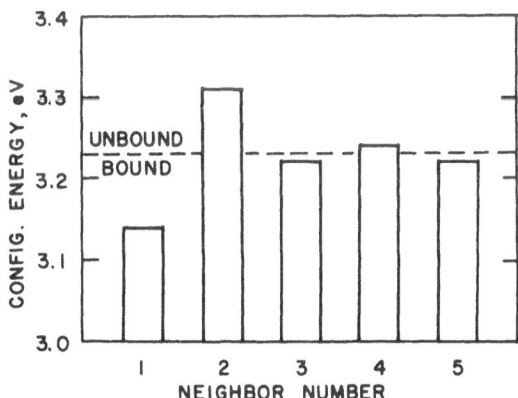

Fig. 4(b). Configuration energy for the Vacancy-LIA complex in α-Fe.

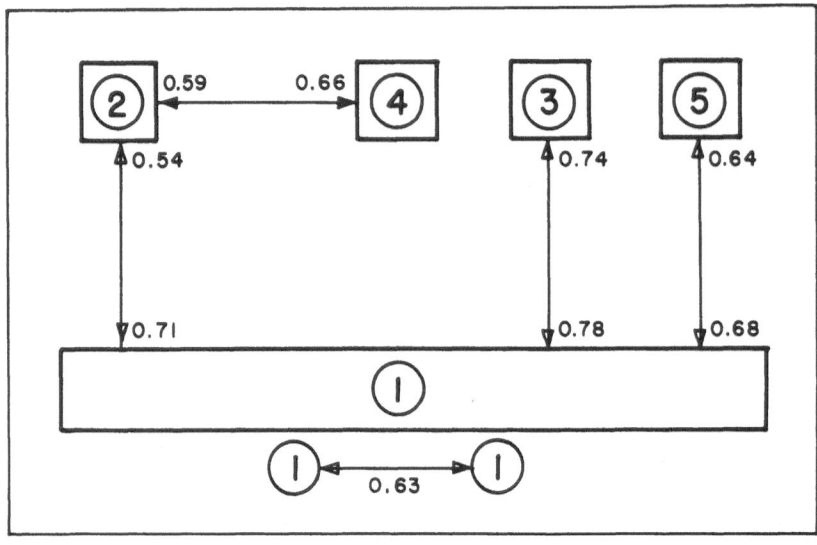

Fig. 5(a). Activation energy required to change from one configuration to another for neighbor indices n = 1 to 5 -- Small Impurity-Vacancy Complex

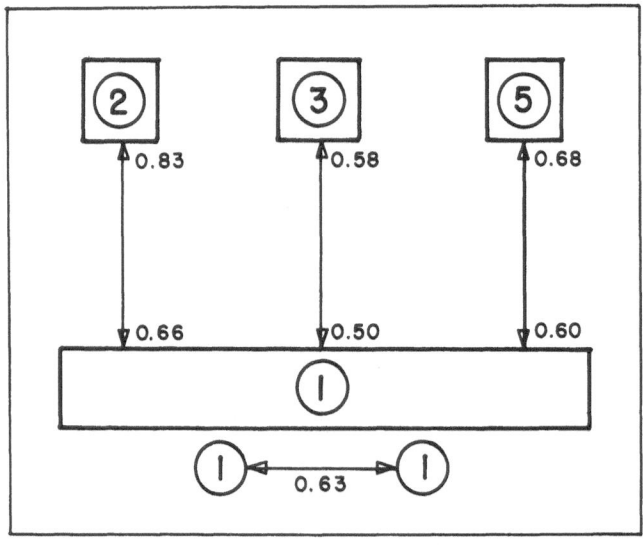

Fig. 5(b). Activation energy required to change from one configuration to another for neighbor indices n = 1 to 5 -- Large Impurity-Vacancy Complex

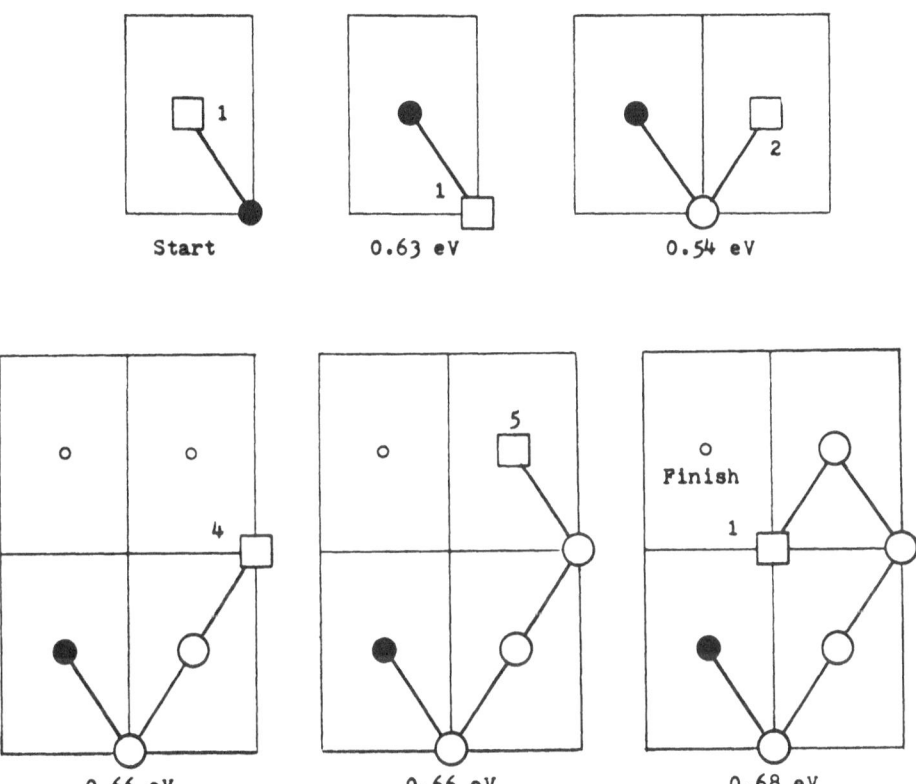

Fig. 6.   Small impurity atom migration jump sequence in α-Fe

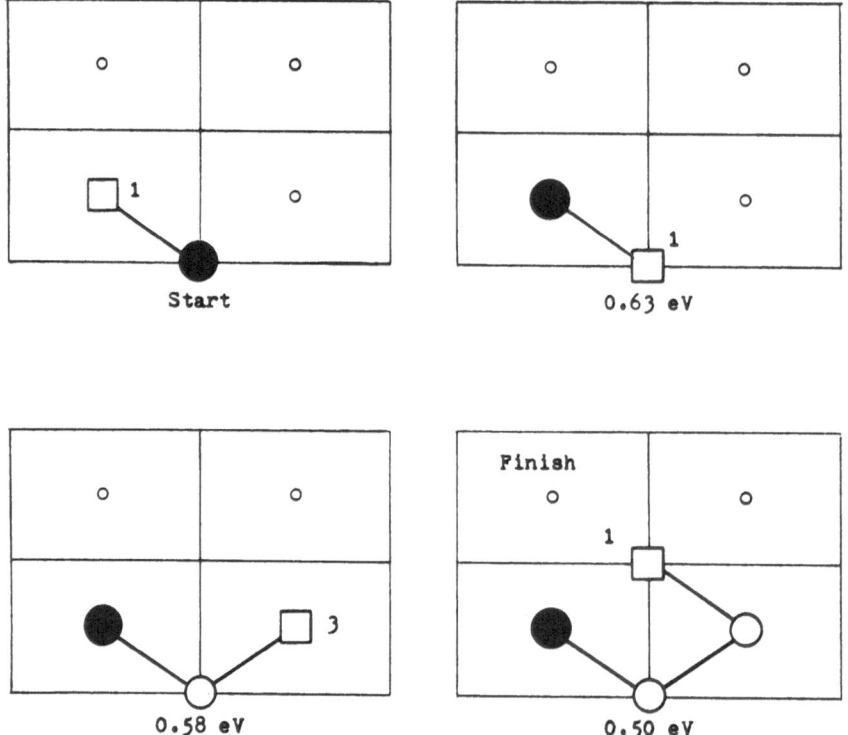

Fig. 7. Large impurity atom migration jump sequence in α-Fe

Table VIII. Configuration Energy for a Substitutional He
Atom Placed Inward from a (110) Free Surface in α-Fe

Substitutional Site	Configuration Energy, eV	Run Number
1 plane inward*	4.325	NC-1773
2 planes inward*	4.359	NC-1775
3 planes inward*	4.368	NC-1774
Bulk position	4.367	NC-2003

*Relative to the (110) free surface plane

'substitutional' He is a vacancy-He complex with the He atom centered in the vacancy. The vacancy configuration energy is 2.89 eV
and that of substitutional He is 4.37 eV. Either injected He (alpha
particle bombardment) or transmutation product He usually adopts an
interstitial position, in the beginning. The configuration energy
for this configuration is 4.86 eV. Hence, the binding energy of a
He-vacancy complex (substitutional He), relative to dissociation
into a vacancy and octahedral He, is

BE(V-He) = (2.89 + 4.86) eV - 4.37 eV

= 3.38 ev.

This constitutes a strong trap for He (or a vacancy).

The Fe-He interaction is wholly repulsive. Because of this,
vacancy clusters serve as strong He atom traps by providing a buf-
fer zone of empty space about the He atom, which otherwise would be
a center of compression. The maximum trapping energy per He atom,
therefore, is equal to the octahedral He configuration energy (4.86
eV). Table IX lists a sequence of vacancy-He complexes. The configuration energy contributed by Fe-He interactions per se appears
in the third column headed, "Energy per He Atom". This quantity
was obtained by subtracting the vacancy cluster (V_n) configuration
energy contribution from the configuration of the entire complex.
A single He atom contributes no more than 0.15 eV to the configuration energy in complexes with six or more vacancies. A $V_{15}(110)$-
faceted cluster can accommodate four He atoms with a configuration
energy contribution of about 0.1 eV per He atom. The data in
Table IX provide reference energies for discussing He atom stabilization of void nuclei.

Four interstitial He configurations were examined: (1) octahedral, (2) tetrahedral, (3) [110]-hybrid split interstitial, and

Table IX. Energy Table for Vacancy Cluster-He Atom
Complexes in α-Fe

Complex	Configura-tion Energy, eV	Energy per He atom eV	Vacancy Cluster BE	Helium to Cluster BE	Run NC-
V_1-He	4.367	1.437	--	3.385 ev	2003
$V_2(1)$-He	6.826	1.169	0.131 eV	3.689 eV	2007
$V_2(2)$-He	7.062	1.469	0.195 eV	3.389 eV	2008
$V_3(112)$-He	9.089	0.902	0.495 eV	3.956 eV	2010
$V_6(Oct.)$-He	15.251	0.151	2.332 eV	4.639 eV	2005
$V_{15}(Oct.)$-He	34.467	0.0	8.943 eV	4.858 eV	1882
$V_{15}(Oct.)$-2He	34.573	0.053	8.943 eV	9.610 eV	2000
$V_{15}(Oct.)$-4He	34.909	0.110	8.943 eV	18.990 eV	2001

Binding energies for He were computed relative to octahedral He.

(4) [111]-hybrid split interstitial. As mentioned previously, a
hybrid split interstitial configuration is one which contains an
impurity atom and a normal atom at its core center, rather than two
normal atoms. The octahedral configuration exhibited the lowest
configuration energy, as is shown in Table X. The tetrahedral configuration is the saddle point configuration for octahedral He
migration in bcc-Fe. By convention this is called the octahedral-
tetrahedral-octahedral (O-T-O) process.

     Simulations were made to study the effect of a (110) void facet
on He migration. An octahedral He atom migrating via the standard
O-T-O migration process was drawn toward a (110) void facet from the
third, second, and first (110) planes inward from the (110) void
facet. Table XI summarizes configuration energies for an octahedral
He atom at positions near a (110) void facet in α-Fe. He was not
retained between the void facet plane and that adjacent to it, nor
was it retained in an octahedral site located in the first (110)
plane inward from the void facet. In each of these two instances,
the He atom spontaneously moved into the void.

Table X. Configuration Energy for He Interstitial
Configurations in α-Fe

Interstitial Type	Configuration Energy, eV	Run NC–
Octahedral	4.858	2004
Tetrahedral	4.889	2009
[111]-Split	5.390	2006
[110]-Split	Unstable*	1861

*Spontaneously transforms to the octahedral configuration.

Table XI. Configuration Energy for Octahedral He
Placed Inward from a (110) Free Surface in α-Fe

Octahedral Site Position	Configuration Energy, eV	Run NC–
1 Plane inward	Unstable*	1766
3/2 Planes inward	4.742	1767
2 Planes inward	4.835	1768
3 Planes inward	4.85**	--
Bulk position	4.858	2004

*Spontaneously moves into the void cavity.
**Extrapolated value

A vacancy which combines with an octahedral He atom to form
substitutional He does not become totally immune to annihilation by
a normal split interstitial in α-Fe. An indirect process can occur
in which Sub. (He) and a normal [110]-split interstitial interact
to form Octa. (He) and to annihilate the vacancy. This process involves the formation and decay of a [111]-hybrid split interstitial.
The energy balance for the process is as follows:

Sub.(He) + [110]-Split = [111]-Hy. Split + Normal Atom + 2.09 eV
(4.37 eV)   (3.11 eV)   (5.39 ev)   (0 eV)

[111]-Hy. Split = Octa.(He) + Normal Atom + 0.53 eV
(5.39 eV)   (4.86 eV)   (0 eV)   .

The activation energy for this process lies between 0.1 and 0.3 eV,
and the total energy release is (2.09 + 0.53) eV = 2.62 eV. Formation

of the metastable [111]-Hy. Split configuration, as an intermediate
step, makes the process kinetically possible. This process is a
good example of how impurity atoms can act both as trapping centers
and as defect annihilation agents.

Johnson's additivity rule works well for vacancy-He complexes.
This statement holds provided the binding energies are computed
relative to dissociation into substitutional He and isolated vacancies. Recall that the binding energies given in Table IX are computed relative to dissociation into octahedral He and vacancy clusters. Strain tensors for octahedral He and the [111]-hybrid split
interstitial are given in Table XII.

Table XII. Strain Tensor Components for Tetragonal He Defects in
α-Fe. ε_i is the i-th strain component and u_i is the associated unit
vector component for the strain coordinate system. $\vec{a}_d$ is the defect
axis.

Defect	i	ε_i	u_i	Defect	i	ε_i	u_i
Octa. Int.	1	0.284	$[100]$	Hybrid Split Int.	1	0.30	$[111]/3^{\frac{1}{2}}$
	2	0.037	$[010]$		2	0	$[\bar{1}\bar{1}2]/6^{\frac{1}{2}}$
	3	0.037	$[001]$		3	0	$[1\bar{1}0]/2^{\frac{1}{2}}$
	$\vec{a}_d = [100]$				$\vec{a}_d = [111]/3^{\frac{1}{2}}$		

5. CARBON IN ALPHA-IRON

Three C-atom defects are especially important in α-Fe. These
defects are (1) the vacancy-C complex, (2) the self-interstitial-C
complex, and (3) C at an octahedral interstitial site. Each of
these basic defects can stabilize an ordered defect structure in
α-Fe.

The vacancy-C complex with the lowest configuration energy is
described in Fig. 8. The binding energy of this complex is 0.42 eV.
When several C atoms are situated near a second-neighbor vacancy
pair, $V_2(2)$, as in Fig. 9, the binding energy per C atom is also
0.42 eV for up to five C atoms at each vacancy. A total of 10 C
atoms can be collected at each $V_2(2)$. This complex lies in a (001)

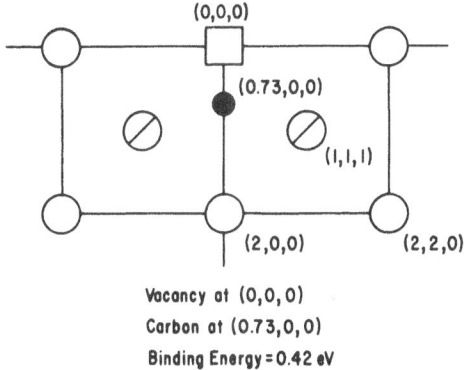

Vacancy at (0,0,0)
Carbon at (0.73,0,0)
Binding Energy = 0.42 eV

Fig. 8.   Vacancy-C complex in α-Fe.

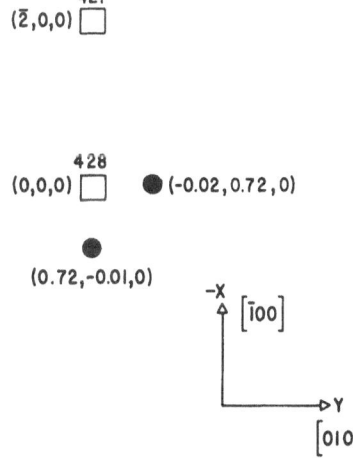

Fig. 9. $V_2(2)$ and two C atoms; binding energy = 1.03 eV

plane and serves as the building module for a defect structure of
indefinite extent on a (001) habit plane. C complexes with a [110]-
split interstitial are shown in Fig. 10. The binding energy for the
I-C complex in Fig. 10(a) is 0.56 eV and that for the I-2C complex
in Fig. 10(b) is 0.76 eV. Here again, C atoms can stabilize a
defect structure. The normal self-interstitial cluster in α-Fe is
a (110) platelet cluster which converts to a dislocation loop at
about 16 interstitials. The additional binding furnished by the
presence of C atoms allows the formation of a (110) cluster of I-C
complexes of indefinite extent. This stabilized structure will not
convert to a dislocation loop. In this structure the binding energy
per complex is about 2 eV, in a normal (110) interstitial cluster
it is about 1 eV per interstitial. The binding energy results for
vacancy-C and interstitial-C complexes are summarized in Table XIII.

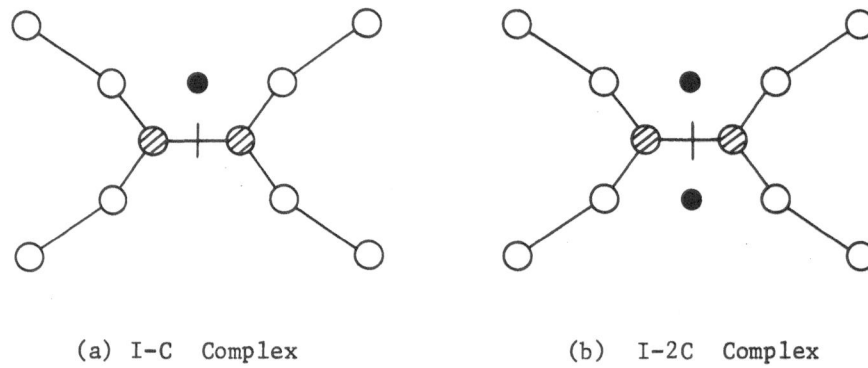

(a) I-C Complex   (b) I-2C Complex

Fig. 10. C complexes with a [110]-split interstitial

Table XIII. Binding Energy for the Basic Carbon-Point-
Defect Complexes in α-Fe.

Complex	Binding Energy, eV
$C-V_1$	0.42
$C-V_2(2)$	0.62
$mC-V_2(2)$	$(0.20 + 0.42\ m)$*
C-I	0.56
2C-I	0.76

*m = 1, 2, ... , 10

The geometry of the octahedral C configuration is described in Fig. 11 and Table XIV. The octahedral C strain tensor is also given in Table XIV. The distance between the C atom and each Fe atom along the tetragonal axis is 1.75 Å (see Table XIV). The configuration energy would be lowered if this distance could be extended to 1.94 Å, which is the separation distance for the most negative Fe-C atom pair energy. Near a free surface (void facet) an extension of the tetragonal distance can be established via surface atom relaxation and C atoms can be strongly bound to a surface. Two guiding principles were observed in the C-free-surface computer experiments: (1) the lowest-energy configuration for a given number of C atoms tends to be that for which the ring distance is least disturbed, while at the same time, the tetragonal distance is lengthened toward 1.94 Å; and (2) the lowest-energy configuration for a given number of C atoms tends to be that in which the surface atom displacement vectors are normal to the free surface. These principles hold for both (110) and (001) free surfaces.

Table XIV. Characteristics of an Octahedral Interstitial C Atom in α-Fe

Carbon atom energy	-1.842 eV
Configuration energy	-1.31 eV
Tetragonal distance	1.75 Å
Ring distance	1.94 Å
Tetragonal strain	0.22
Ring strain	-0.06

The octahedral C displacement pattern at a (110) surface is described in Fig. 12. This is not the configuration with the largest binding energy per C atom, however, because the surface atom displacement vector (atom C) is not normal to the surface. When a pair of C atoms are introduced, as in Fig. 13, surface atom displacement vectors normal to the surface are obtained and the binding energy increases to 0.44 eV per C atom. All ordered (110) surface C platelets, therefore, will tend to be built up on the basis of a C atom pair module.

Atom pairs are also preferred at a (001) free surface. An ordered structure of indefinite extent could be built at a (001) surface by arranging C atoms as shown in Fig. 14.

Table XV summarizes the binding energy results for C atoms at (001) and (110) free surfaces (void facets) in α-Fe. The binding

Table XV.   C Atom Binding Energies at
(110) and (001) Free Surfaces in α-Fe

Carbon	Binding Energy, eV	
Configuration	(110)FS	(001)FS
Single Carbon	0.26	0.39
Carbon pair	0.88	0.85
Carbon quartet	1.91	1.77

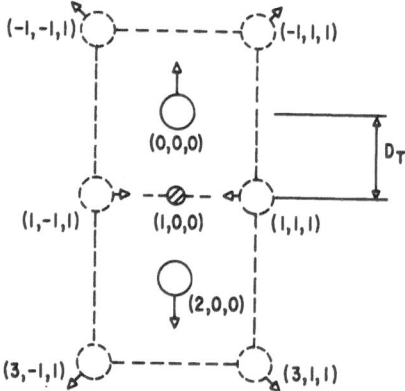

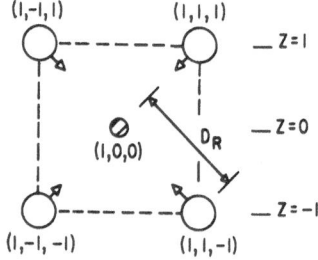

Fig. 11.   Geometry of the octahedral C configuration (see also
Table XIV)

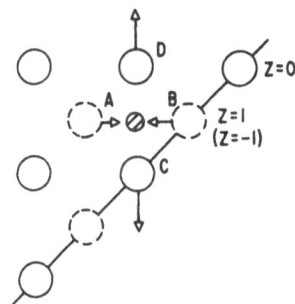

Atom	Strain Vector		
	ΔX	ΔY	ΔZ
A, Z=1	0.039	0.050	-0.047
B, Z=1	-0.004	-0.049	-0.050
C	0.345	0.121	0.0
D	-0.176	0.0	0.0
Carbon	0.089	-0.005	0.0

BE = 0.26 eV

Fig. 12.   Octahedral C displacement pattern at a (110) surface in α-Fe

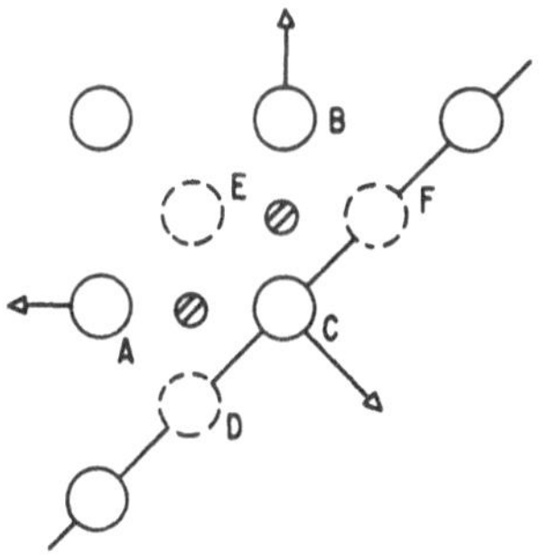

Fig. 13.   Surface atom displacement pattern in α-Fe resulting from the introduction of a _pair_ of C atoms (⊘)

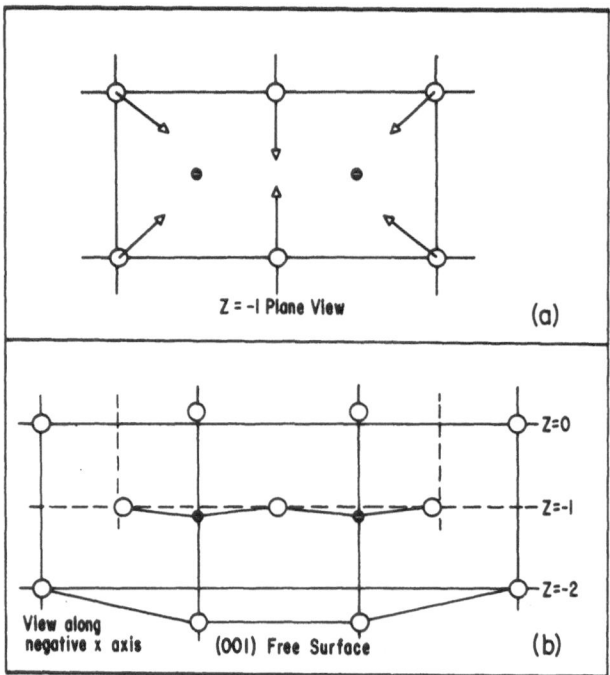

Fig. 14. Positioning of C atoms [⊖] as shown results in an
         ordered structure of indefinite extent.

energy per C atom in a precipitate in bulk α-Fe is 0.31 eV in
Johnson's model [1]. From Table XV, one sees that the binding energy
per C atom at a void facet exceeds that in a precipitate in bulk
α-Fe. This statement is true for platelets of two or more C atoms.
Hence, C atoms ejected from a precipitate in irradiated Fe alloys,
by atomic collision cascades, will preferentially be collected at
void facets rather than returning to a precipitate. A detailed
description of C complexes in fcc, bcc, and hcp crystals can be
found in Ref. [9] and particular C stabilization of void corners and
edges is discussed in Ref. [10].

ACKNOWLEDGMENTS

     The DYNAM program was written by R. H. Evans, Jr. This work
was sponsored by the U.S. Air Force under Contract AFML-TR-70-260.
The impurity atom complex calculations were performed by M. F. Beeler,
Won Pyo Chun, and R. H. Evans, Jr. This computational work was
sponsored by the U.S. Atomic Energy Commission under Contract
AT(40-1)-3912.

REFERENCES

1. JOHNSON, R.A., DIENES, G.J. and DAMASK, A.C., Acta Met. $\underline{12}$
    1215 (1964).
2. JOHNSON, R.A., Phys. Rev. $\underline{134}$ A1329 (1964).
3. WILSON, W.D. and BISSON, C.L., Sandia Laboratories Report
    SCL-DC-70-45, March, 1970.
4. COCHARDT, A.W., SCHOECK, G. and WIEDERSICH, H., Acta Met. $\underline{3}$
    533 (1955).
5. GIBSON, J.B., GOLAND, A.N., MILGRAM, M. and VINEYARD, G.H.,
    Phys. Rev. $\underline{120}$ 1229 (1960).
6. BEELER, J.R., Jr., in Interatomic Potentials and Simulation of
    Lattice Defects, GEHLEN, P.C., BEELER, J.R., Jr. and JAFFEE,
    R.I., eds., Plenum Press, 1972, pp 735-751.
7. JOHNSON, R.A., Phys. Rev. $\underline{145}$, 423 (1966).
8. BEELER, J.R., Jr. and JOHNSON, R.A., Phys. Rev. $\underline{156}$ 677 (1967).
9. BEELER, J.R., Jr., in Interatomic Potentials and Simulation of
    Lattice Defects, GEHLEN, P.C., BEELER, J.R., Jr. and JAFFEE,
    R.I., eds., Plenum Press, 1972, pp 339-374.
10. BEELER, J.R., Jr. and BEELER, M.F., in Effects of Radiation on
    Substructure and Mechanical Properties of Metals and Alloys,
    ASTM STP 529, October, 1973.

PHYSICAL PRINCIPLES OF SOLID SOLUTION STRENGTHENING IN ALLOYS

E. W. Collings

BATTELLE
Columbus Laboratories
Columbus, Ohio 43201

H. L. Gegel

Air Force Materials Laboratory
Wright-Patterson Air Force Base
Ohio 45433

ABSTRACT

An electronic component to solid-solution strengthening can be discussed in terms of a series expansion for the free energy, $F = \sum F_r$. We postulate that the resistance to deformation of a metal can be gauged by the extent to which F is structure-dependent; i.e, by the rapidity with which the above series converges. This convergence is controlled by a parameter $E_{1,2}$, a measure of the solvent-solute atomic potential difference. Recognizing that $E_{1,2}$ is also a measure of the electron-scattering potential of the dissolved impurity ion, it is possible to construct a relative scale of atomic potential differences, and consequently solid-solution-strengthening capacities for various solutes in Ti, by simply measuring specific resistivities; i.e., resistivities per atomic percent solute. In so doing we find that for rapid solid-solution strengthening in Ti alloys we should look to nontransition elements (B-metals) rather than transition elements. Indeed the latter are useful principally as stabilizers of the bcc structure at ordinary temperatures rather than as strengthening agents. Secondly, we note that the principles governing the solid-solution strengthening of Ti by B-metals apply equally well to the strengthening of other related host transition elements and even transition-metal binary alloy hosts. Finally, we discuss a new measure of atomic potential difference--a "reduced interaction strength parameter", I. which for a given solute-solvent pair yields the metallic alloy equivalent of the Pauling electronegativity difference. A scale of equivalent metallic electronegativity values has been constructed from the results of relative

vapor-pressure studies of numerous appropriate binary alloy systems
and the validity of I as a solid-solution strengthening parameter
is demonstrated.

1. INTRODUCTION

Central to any fundamental approach to the mechanical properties
of metallic solids is the concept of electron-moderated interactions
between ions. This may be taken as the basis of lattice stability,
solution hardening, and fracture. Ionic interactions may be
resolved into pair, triplet, etc. components, as indicated by the
following expansion for the free energy, F [1-5]

$$F = F_e + F_c + F_{si} \quad , \tag{1}$$

where F_e = the free energy of the electron gas plus the self-
         energies of the ions surrounded by their local
         screening charges,

F_c = free energy of interaction of the electron gas with
         the ions individually,

F_{si} = $F_{s2} + F_{s3} + \ldots$ for the sum of the pair-wise interac-
         tions between the ions, plus the sum of triplet
         interactions, plus etc.

1.1 Pure Metals

All the structure dependence of F resides in the series F_{si},
the rapidity of convergence of which depends on the strengths of
the atomic potentials expressed in some suitable form. If the
potentials are weak, a termination of F_{si} after the first term, F_{s2},
is permissible. This approximation, which is equivalent to using
second-order perturbation theory [6-8] leaves F only slightly structure dependent. The corresponding mechanical characteristic
(exhibited by weak-potential metals such as Na and K) is softness.

As the strengths of the atomic potentials of a sequence of
metals increase, more terms in F_{si} must be retained, F_{si}/F increases,
and F becomes more and more structure dependent. As a result,
these metals fall into a sequence of increasing hardness. Cohen [1]
has suggested the use of E_g/Δ as a suitable measure of atomic potential strength where E_g is the width of a low-lying band gap at the
zone boundary, and Δ is the width of the energy band itself.

1.2 Alloys

The above argument carries over to alloys if E_g is replaced by
$E_{1,2}$, the difference in potentials between solute and solvent atoms.
Thus departures from (a) second-order perturbation theory for alloys,

and (b) the validity of the rigid-band model, go hand-in-hand with increasing solution hardening. Since electron-impurity scattering also increases with $E_{1,2}$, we expect that the relative "specific electrical resistivity" (i.e., resistivity per atomic percent solute) should afford a useful guide to the relative hardening capacity of a given class of solute.

1.3 Electrical Resistivity of an Alloy

The electrical resistivity, ρ, of an alloy may be written

$$\rho = \rho_{impurity} + \rho_{phonon} \quad , \tag{2}$$

where ρ/unit concentration (at.%) is the specific resistivity referred to above. Under Mattheissen's rule ρ_{phonon} would represent the solute-independent, temperature-dependent (so-called "ideal") resistivity of the host. In the more likely event that the presence of the solute perturbs the phonon spectrum of the host, we must amend Eqn. (2) and write

$$\rho = \rho_{impurity} + (\rho_{impurity} + \rho_{ideal})_{phonon} \quad . \tag{3}$$

According to Eqn. (3), the impurity may produce a local scattering of electrons ($\rho_{impurity}$), or it may perturb the phonon spectrum to such an extent that additional electron-phonon scattering occurs as a bulk effect, leading to the contribution ($\rho_{impurity} + \rho_{ideal})_{phonon}$. The modified phonon spectrum is related in some alloy systems to lattice stability. Whether or not impurity scattering is dominant depends on $E_{1,2}$, and hence on specific electrical resistivity.

2. BINARY TITANIUM-BASE ALLOYS

As shown in Fig. 1, Ti-base alloys can be divided into two classes depending on the composition-dependences of their electrical resistivities. The symbol B is used to represent a non-transition element, typically a B subgroup metal such as Al, Ga, Sn, etc., but it may also represent a nonmetallic element such as N, O, C, etc. T_2 represents a transition element "near" titanium in the transition-metal block of the periodic table. In the light of the preceding remarks, the figure predicts that the solid-solution strengthening capacities of B-metals in Ti will be significantly greater than those of nearby transition elements. This is confirmed in Fig. 2 which compares the strengthening of Ti, in terms of the proportional limit,

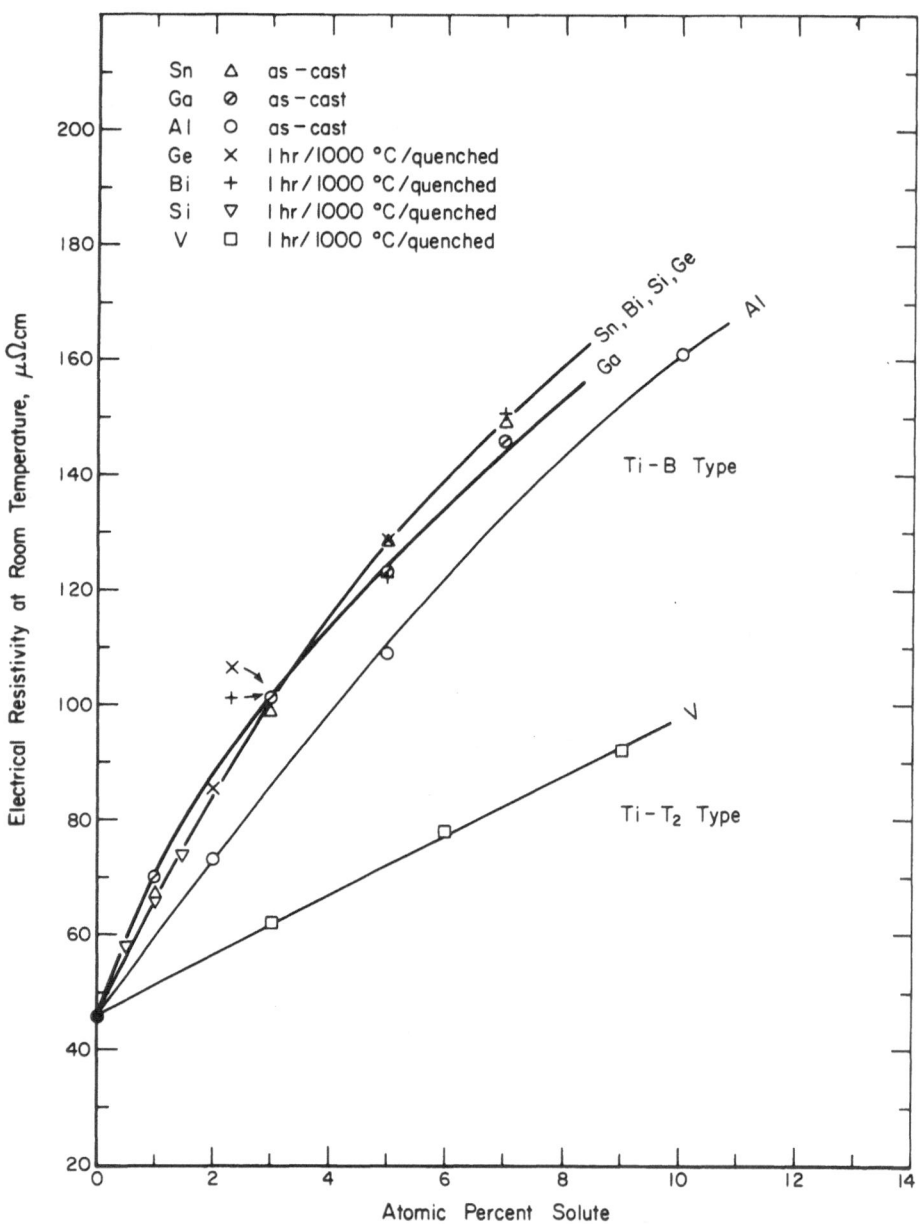

Fig. 1. Room temperature electrical resistivities of two classes
of titanium alloys

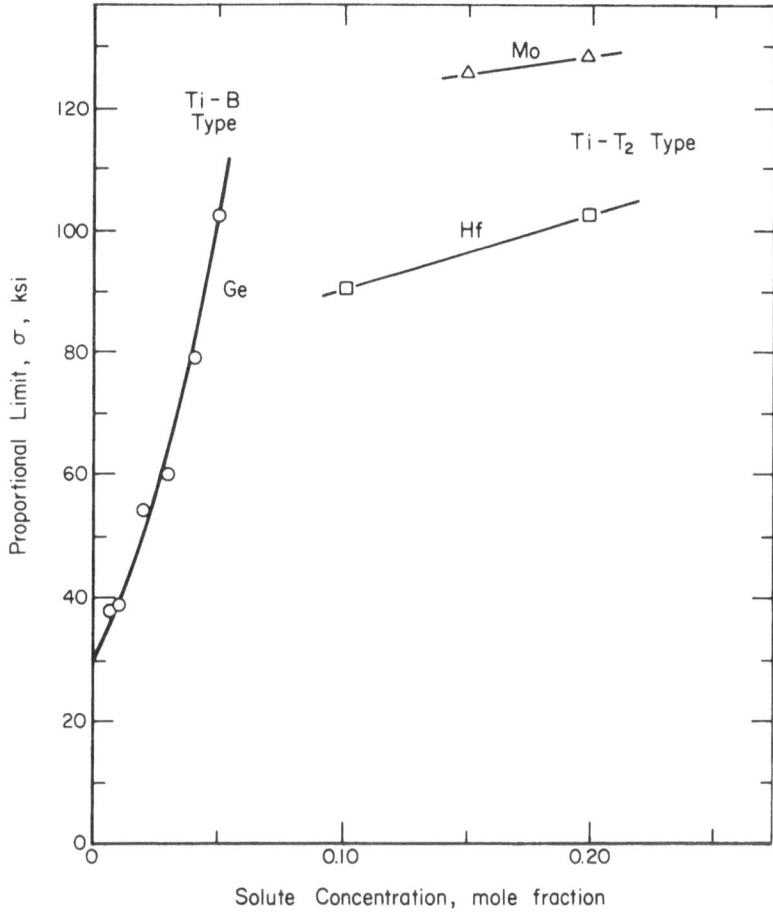

Fig. 2. Compressive strengths in terms of the proportional limit, σ, for two classes of titanium alloys

σ, by a representative B-metal (Ge) with that produced by a pair of representative transition elements (Hf and Mo).

Ti-B and Ti-T_2 clearly represent two distinct classes of Ti-base alloys. The former are characterized by large specific electrical resistivities and moderately rapid rates of solid-solution strengthening. The strengthening mechanism, which has to do with short-range solute-solvent bonding, is to be discussed in more detail later. The crystal structure of Ti-B alloys is hexagonal close packed (hcp), similar to that of the low-temperature (<885°C)

allotrope of pure Ti. On the other hand, the Ti-T_2 alloys are body-
centered cubic (bcc). B-elements such as Sn, Ga, Al, O, and N are
called α-phase stabilizing elements, since they stabilize the low
temperature α (i.e., hcp) phase to temperatures in the vicinity of,
or above, the allotropic transformation temperature of pure Ti;
whereas the transition elements such as V, Mo, etc. stabilize the
high temperature β (bcc) phase to room temperature. The principal
strengthening mechanism in the Ti-T_2 alloys can be thought of as a
progressive isothermal stabilization of the β phase with increasing
solute concentration.

3. Ti-B-TYPE BINARY ALLOYS

A good example of a Ti-B-type alloy is Ti-Al. At low solute
concentrations (<10-15 at.% Al) the rapidly quenched alloy is disordered. As the Al concentration is further increased, there is a
tendency for ordering to take place, to atomic arrangements ($α_2$)
based on the Ti_3Al structure as in Fig. 3(a). The ordering is
accompanied by a drop in Fermi density-of-states properties (e.g.,
magnetic susceptibility), as shown in Fig. 3(b), as electrons tend
to become localized or more tightly bound. Strong evidence for
directional bonding is to be found in the result of Gehlen's [9]
X-ray short-range-order (SRO) investigation, which indicated that
the Ti atoms appeared to be displaced away from their expected positions in the DO_{19} lattice. This could be interpreted in terms of a
crowding of charge along the Ti-Al directions (Fig. 3(a)), but other
explanations in terms of Ti-Ti bonding are also admissible.

The strengthening effect of Al in Ti, even in the low-concen-
tration range of 0-4 at.% is related to the electronic bonding
characteristics of the ordered intermetallic compound. We suggest
that, in the vicinity of a dissolved Al atom, the same kind of electronic bonding is set up as appears in the compound. Strengthening
is a result of the strong pinning which occurs at the site of a dissolved Al atom, when the locally bound matrix atoms find themselves
part of a passing dislocation.

4. Ti-T_2-TYPE BINARY ALLOYS

4.1 Lattice Stability

As indicated earlier, transition elements confer strengthening
to Ti through bcc lattice stabilization. This is a bulk, or elastic
modulus, type of mechanism, in contrast to the local dislocation-
pinning mechanism characteristic of B-element strengthening. Fig. 4
presents as a function of composition some useful electron-phonon,

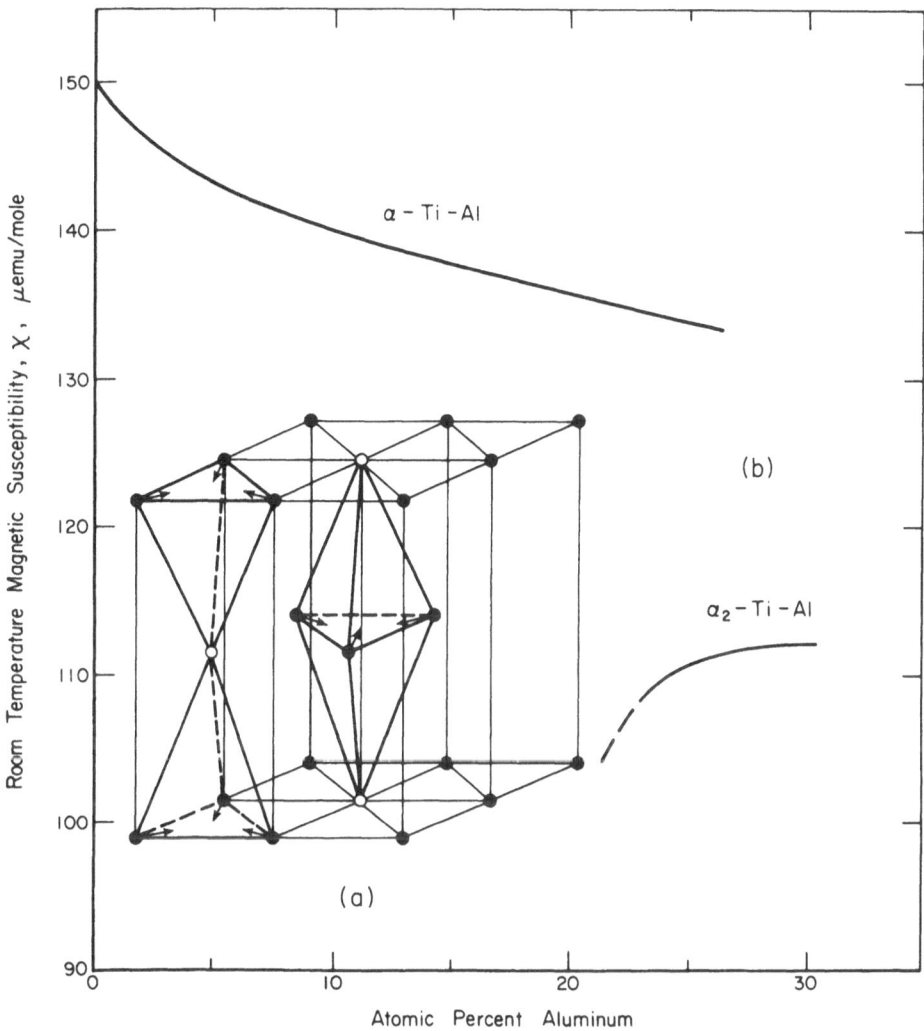

Fig. 3. Bonding and density of states properties of Ti-Al binary alloys: (a) Structure of ordered Ti_3Al indicating effects of directional bonding (a); (b) Room temperature magnetic susceptibility of Ti-Al alloys indicating a pronounced drop in Fermi density of states upon entering the ordering regime.

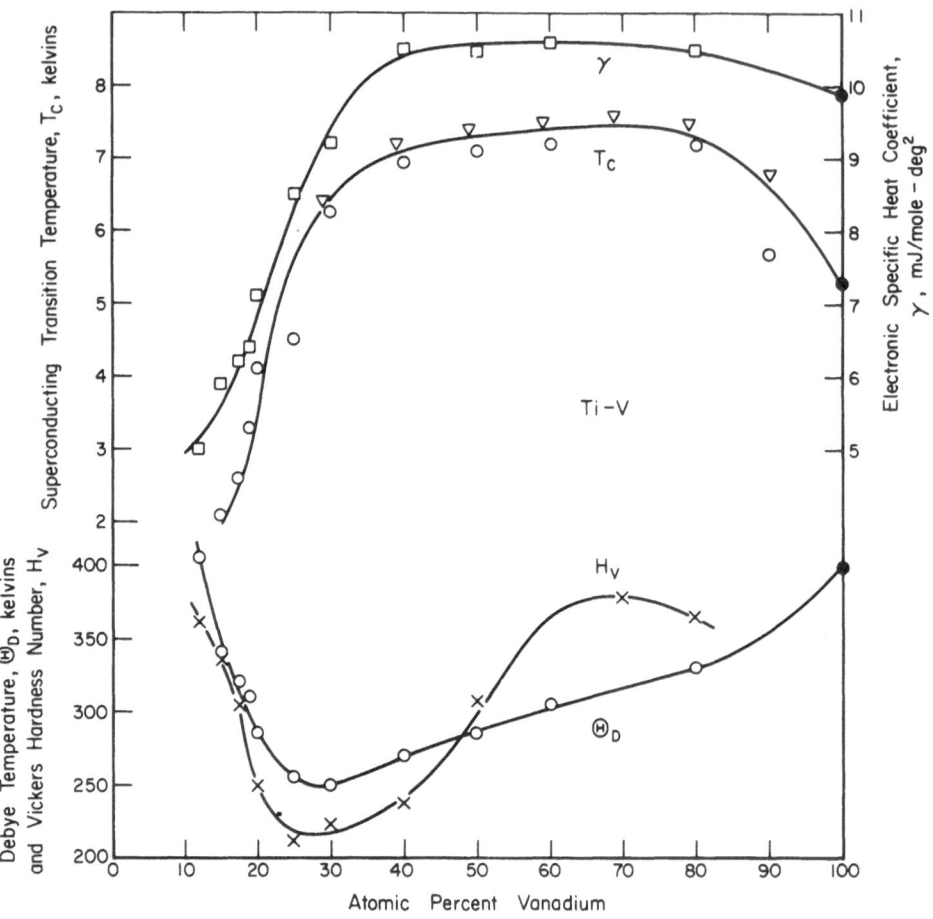

Fig. 4. Comparative behavior of electronic specific heat coefficient, γ, superconducting transition temperature, T_c, Debye temperature, Θ_D, and Vickers hardness number, H_v, as functions of composition in the Ti-V system

elastic, and mechanical properties of a representative transition-metal-binary alloy, Ti-V. Reading from right to left, as Ti is added to V the destabilization of the bcc lattice results in a mechanical softening (decrease in hardness, H_v), a lattice softening (reduction of the Debye temperature, Θ_D), and an increase in the superconducting transition temperature, T_c. The turning points in the curves are induced by ω-phase precipitation, a product of bcc lattice instability as the martensitic transformation (bcc $\underrightarrow{\text{spontaneous}}$ hcp) composition is approached. The bcc lattice transforms athermally by a displacive mechanism during cooling to α' (martensite) when the vanadium concentration is sufficiently low as to yield an average electron/atom ($\textbf{3}$) ratio of about 4.1. The inverse relationship between Θ_D and T_c is a quite general property of transition-metal-binary alloys in the average group-number ($\textbf{3} \equiv$ electron/atom ratio) range of 4 to 6. Table I suggests that T_c makes a useful index of lattice softening, and that in the "bcc field" bulk stiffening accompanies the addition of a transition metal to Ti in the range 4.4 $\gtrsim \textbf{3}$ <6. The shear modulus data in Fig. 5 also indicate the existence of a universal (for transition metals) relationship between bcc lattice stability and $\textbf{3}$. Fig. 5 also indicates that an increased modulus-type (bulk) solution strengthening accompanies any such alloying that results in an increase in $\textbf{3}$, provided that the value $\textbf{3} \sim 6$ is not exceeded.

Table I. Superconducting transition temperatures, T_c, and Debye temperatures, Θ_D, for Groups IV through VI transition elements and some alloys

	Ti	Ti-V (50 at.%)	V	Cr
Θ_D	425	285	399	630
T_c	0.3	7.1	5.4	--

	Zr	Zr-Nb (50 at.%)	Nb	Mo
Θ_D	290	238	277	460
T_c	0.6	10.3	9.2	0.9

	Hf	Hf-Ta (50 at.%)	Ta	W
Θ_D	252	209	258	390
T_c	< 0.02	6.7	4.5	0.01

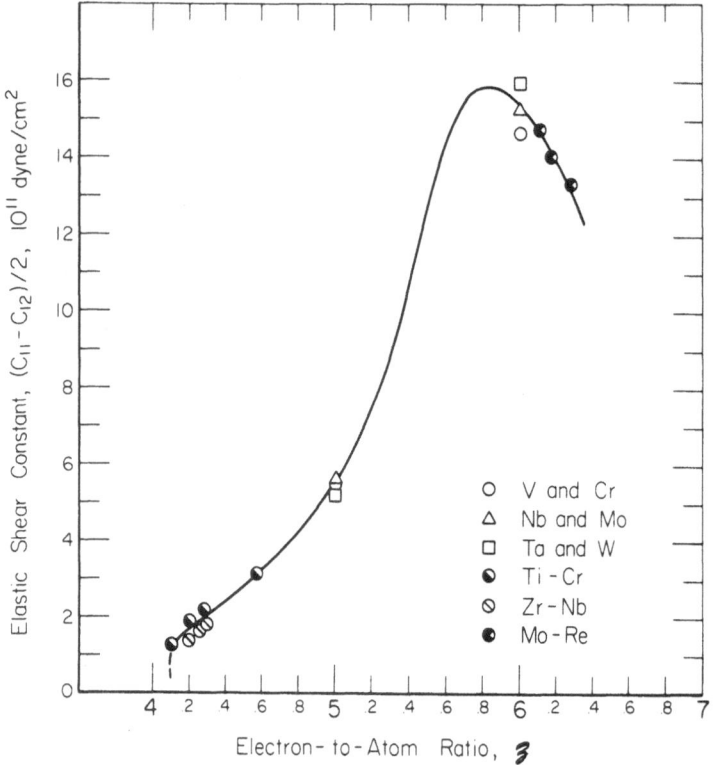

Fig. 5. Relative lattice stabilities of transition metals and their binary alloys in the electron-to-atom ratio range 4.1 to 6.2

4.2 Additional Solution-Strengthening Effects

By studying isoelectronic systems using the superconducting transition temperature, T_c, as an indicator, it can be shown that some increase in lattice stabilization (or strengthening) <u>always</u> accompanies transition-metal-binary alloying (whether $\mathfrak{z}$ increases or decreases). In Fig. 6 we see that T_c in the isoelectronic systems V-Nb and V-Ta <u>always</u> decreases in the early stages of alloying. Again, for $\mathfrak{z}$ = 5.0, Table II compares the pure metal Nb with the "equivalent" alloy Ti-Mo (50 at.%). We see that alloying at constant $\mathfrak{z}$, i.e., at a fixed position in the bcc stability range, stiffens the lattice. This type of strengthening is due principally to atomic size differences, so-called "size-effect". In general it must compete with the more dominant mechanisms of (a) lattice stabilization/destabilization, in Ti-T_2-type systems; or (b) local-interaction mechanisms, as in the Ti-B-type alloys.

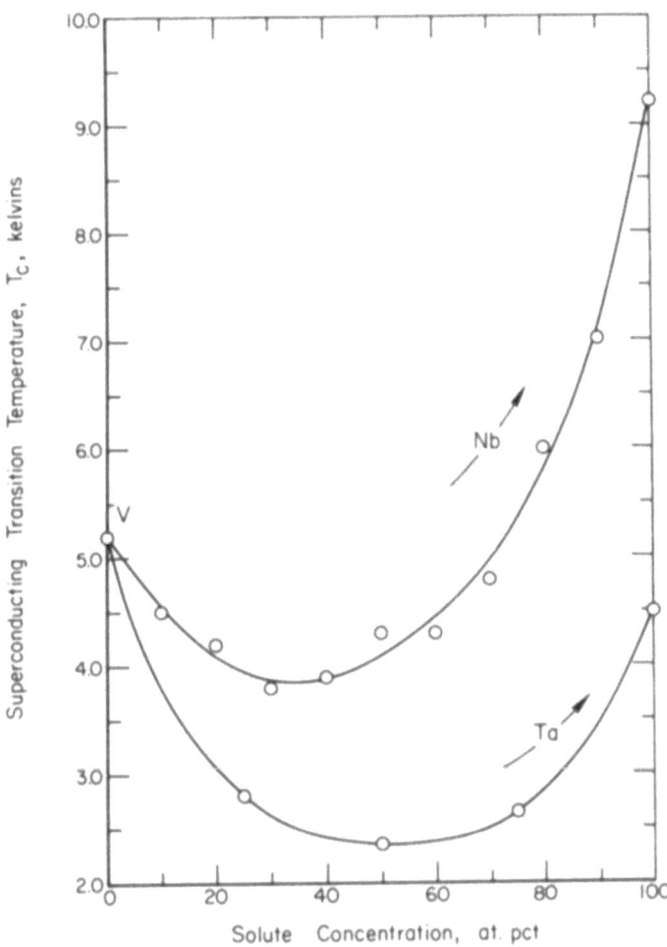

Fig. 6. Variation of superconducting transition temperature for
the alloy systems V-Nb and V-Ta

Table II. Comparison of T_c and Θ_D for the
isoelectronic materials Ti–Mo (50 at.%) and
Nb

	Ti–Mo (50 at.%)	Nb
Θ_D	390	277
T_c	1.6	9.2

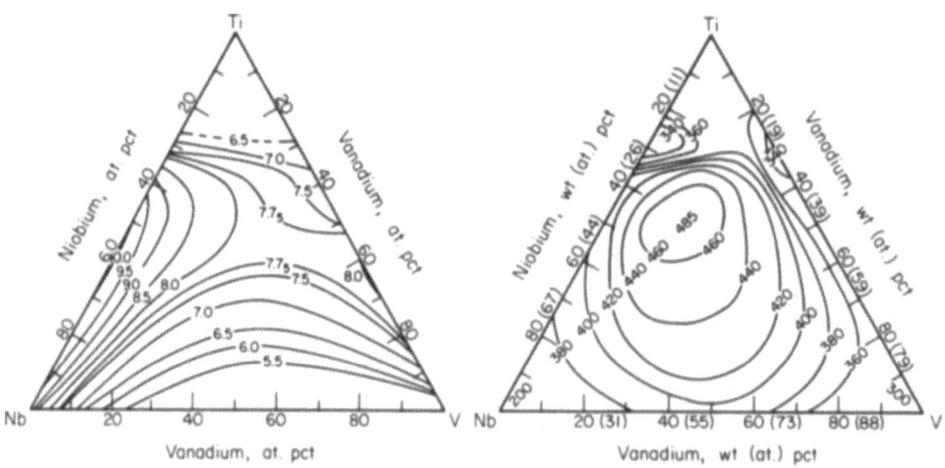

Fig. 7. Relationship between superconducting transition temperature
and hardness in the ternary system Ti–V–Nb

5. TERNARY ALLOYS

5.1 Transition-Metal-Ternary Alloys

Fig. 7, which shows T_c increasing as the hardness H_v decreases,
emphasizes the value of superconducting transition temperature as
an indicator of alloy softening. After recasting the same superconductivity data in the form shown in Fig. 8, it is possible to
discuss, with reference to a single system, two characteristics of
transition-metal alloy strengthening. Firstly, we see a solute-
strengthening effect accompanying the addition of Nb to Ti–V in such

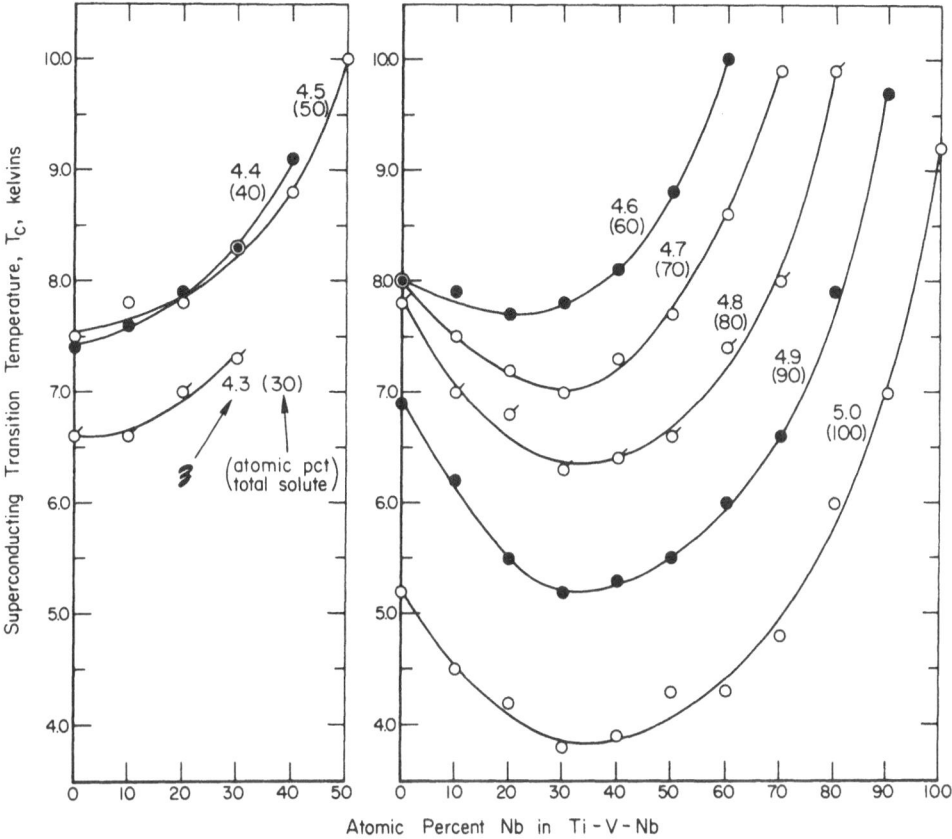

Fig. 8. Isoelectron-to-Atom ratio curves demonstrating the effect of Nb substitution in Ti-V alloys

a way as to maintain $\mathfrak{z}$ constant. Secondly, as $\mathfrak{z}$ decreases from the value 6, the continuous general increase in T_c (until precipitation intervenes at $\mathfrak{z} \sim 4.2$) is indicative of a continuous destabilization-softening of the bcc lattice.

5.2 Ti-T$_2$-B-Type Ternary Alloys

By virtue of the weakness of the Ti-T$_2$ interaction as compared with the Ti-B interaction, a Ti-T$_2$ alloy may be regarded as a "pseudo-transition element" which should be amenable to being solution-strengthened by the addition of B-elements. The mechanical

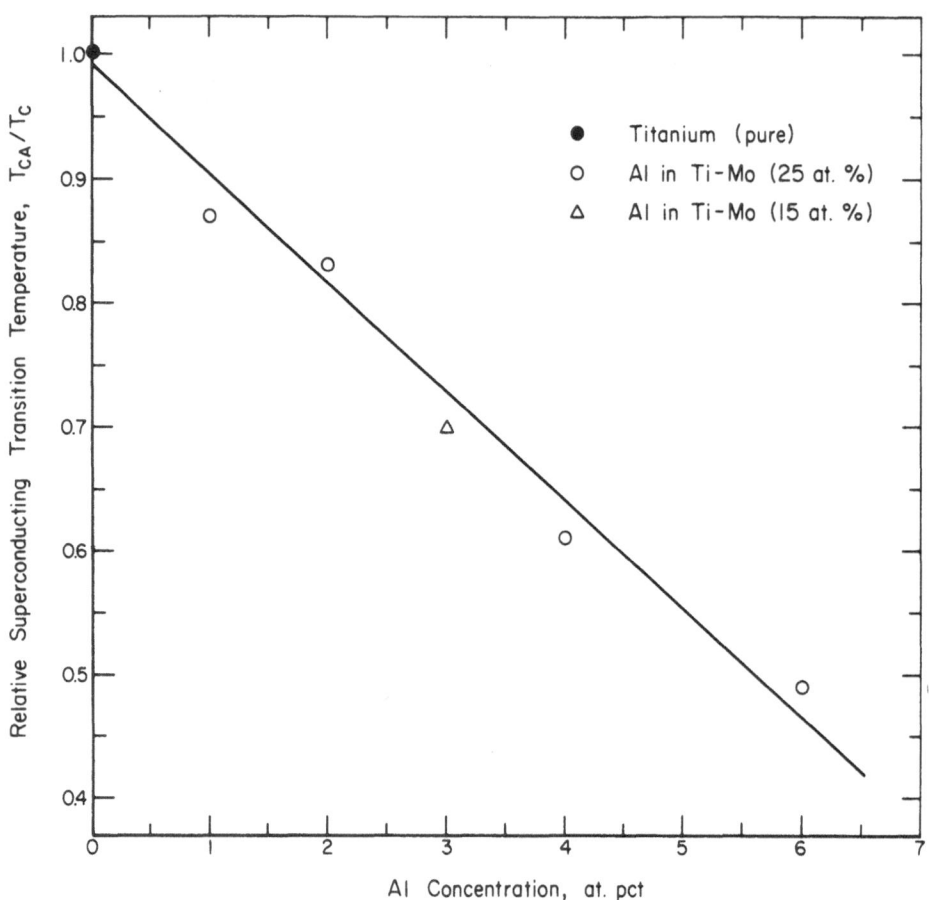

Fig. 9. Depression of superconducting transition temperature in Ti-Mo alloys following solution strengthening by Al

properties of such alloy systems are to be discussed in more detail later; however, at this stage we draw attention to the superconductivity results which indicate the considerable lattice stiffening which accompanies the addition of B-metals or mixtures of B-metals to Ti-Mo (25 at.%). Fig. 9 shows the influence on T_c of B-metal additions to our representative "pseudo-transition metal" base.

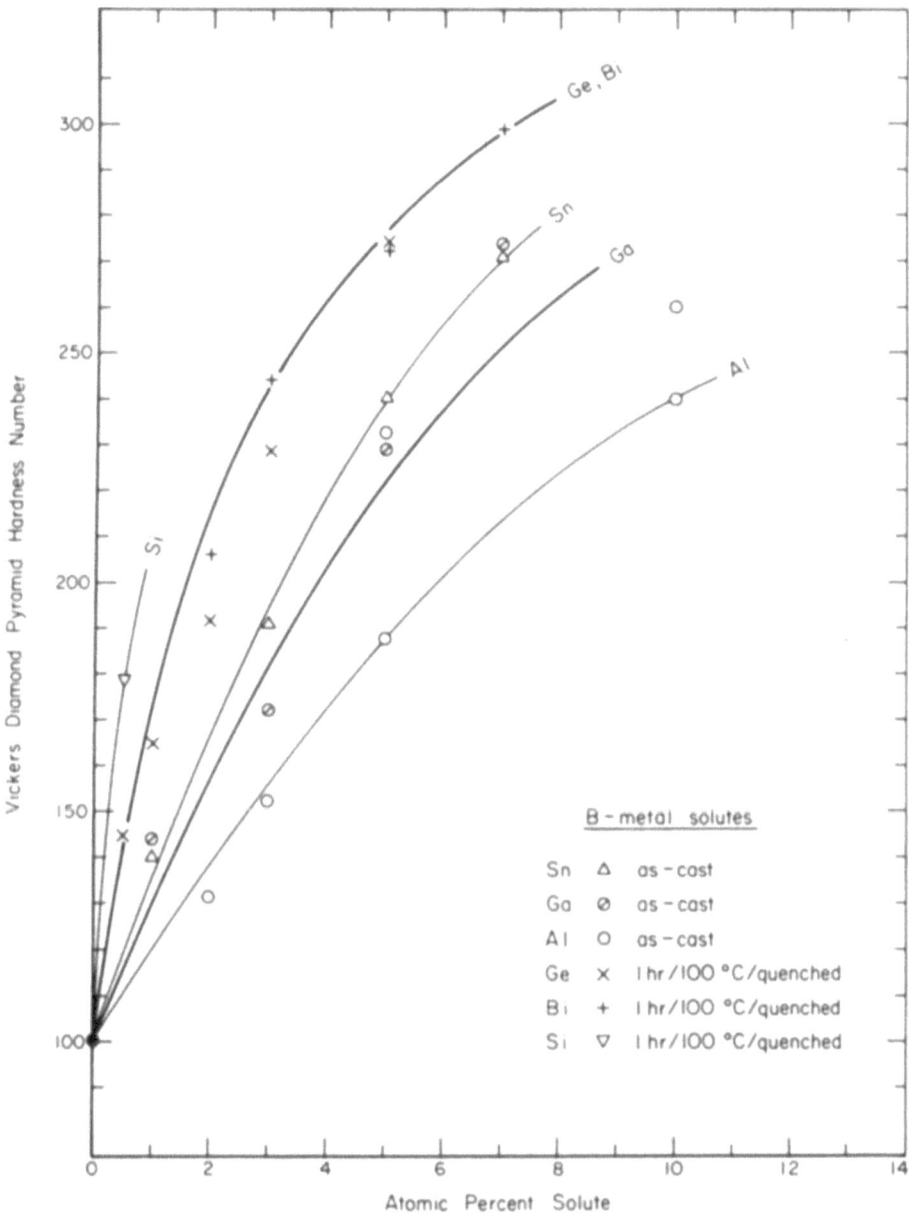

Fig. 10. Vickers hardness <u>versus</u> composition for six Ti-B-type alloys

6. THERMODYNAMICS AND STRENGTHENING WITH RESPECT TO
    VARIOUS NONTRANSITION AND TRANSITION-METAL SOLUTES
    IN TITANIUM

In earlier sections, we have indicated the use and significance
of electrical resistivity as an index of solid-solution strengthening capacity. Returning to Fig. 1, we note that although a clear
separation between Ti-B and Ti-T_2 alloy classes is possible, it is
difficult in that figure to distinguish among the specific resistivities of the B-metals themselves. The results of a recently
completed set of hardness measurements on Ti-B alloys are shown in
Fig. 10. The hardness data indicate that dissolved B-metals differ
significantly in strengthening capacity, with for example a sequence
Si > Ge, Bi > Sn > Ga > Al being clearly discernible. For use in
basic alloy design, strengthening indices capable of unambiguously
distinguishing between the properties of different B-metals are
desirable. With the aid of a regular-solution thermodynamic interaction parameter, Ω_{ij}, such a scale can be devised. The significance
of the interaction parameter, and its eventual use in deriving an
equivalent "electronegativity scale" for alloy situations, will be
discussed in detail in subsequent sections. It is sufficient here
to indicate (Fig. 11) that Ω_{ij} not only separates B-metal and
transition-metal solutes into two regimes differing in sign (the
negative sign for Ti-B indicating short-range-ordering, and the
positive sign for Ti-T_2 indicating clustering); but in addition
provides a substantial separation of elements within each regime.
It is, for example, to be noted that the Ω_{ij} values for the B-metals
referred to above increase in the sequence Al < Ga < Sn.

6.1 The Regular-Solution Thermodynamic
         Interaction Parameter, Ω_{ij}

Solid-solution strengthening is intimately related to factors
which determine the compositional range and stability of a binary
alloy. The following factors seem to be important

(a) An electronegativity difference, $\varepsilon_i - \varepsilon_j$, between
     the pairs of elements

(b) An atomic size difference

(c) A lattice stability parameter, such as, for example,
     the shear modulus $C' = (C_{11} - C_{12})/2$.

Alloys of titanium can be broadly divided into two classes
based on thermodynamic or electronegativity difference considerations. The pairwise interaction parameter, Ω_{ij}, from regular solution theory [10] divides Ti-base alloys into Ti-B-type alloys
(B = Al, Ga, Sn, etc.) which have strong negative pairwise inter-

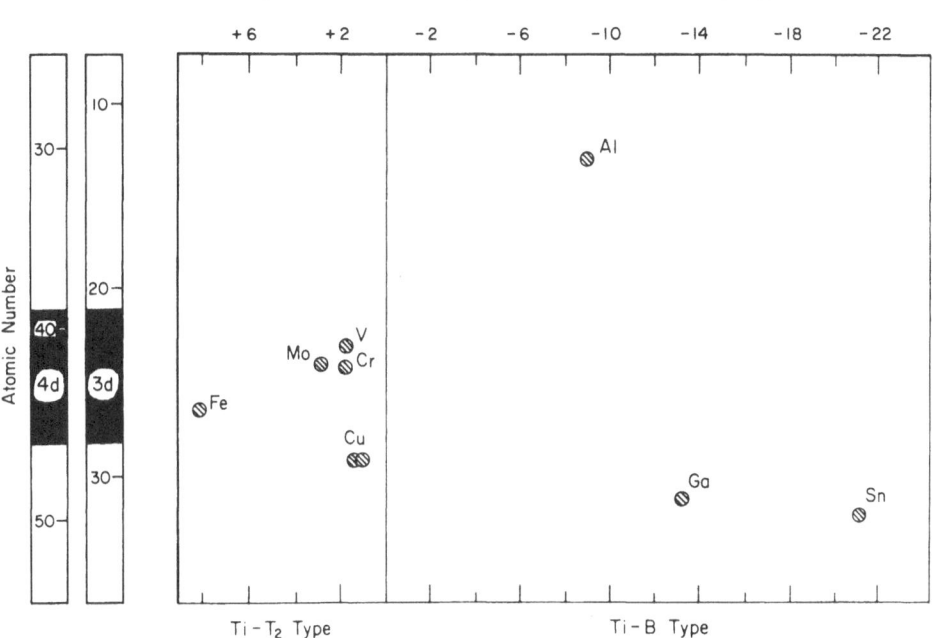

Fig. 11. Thermodynamic evidence for the existence of two classes of Ti-base alloys

actions and into Ti-T_2-type alloys (where T_2 is another transition element) having weak positive pairwise interactions. The relative magnitudes of the pairwise interaction parameters for the two alloy classes are of the order 10:1 (comparing Sn with V).

In order to develop a model for solid-solution strengthening and to assess the solution-strengthening potency of a given alloying element, it would be useful to develop some kind of electronegativity scale appropriate for metallic alloying. The Ω_{ij} values may offer such a possibility and will be used to develop a quantitative scale comparable in many respects to the Pauling electronegativity scale. The principal difference between the two electronegativity scales has to do with the energy zero. In this case, Ti was used as the solvent, or reference metal. It is not known whether this new scale can be universally applied or not, but one can, for example, use the Ti-based data to calculate the Ω_{ij}'s for Ga-Al and V-Cr, obtaining values which agree with the above experimental values [11,12]. The basis for the new scale will be briefly outlined, while the experimental details for obtaining Ω_{ij} values are published elsewhere [13-16].

The free energy difference between an element, i, in solution and in its pure form is given by

$$\overline{G}_i - \overline{G}_i^0 = RT \ln a_i \equiv \Delta G_i^{tr} + RT \ln N_i + \overline{G}_i^{xs} \quad , \tag{4}$$

where ΔG_i^{tr}= free energy difference between i in its standard state and that of the pure element when it has the structure and lattice parameter of the solution. The partial molar excess free energy, $\overline{G}_i^{xs}$, defined in terms of the pairwise interaction parameter, Ω_{ij}, is written as

$$\overline{G}_i^{xs} = N_j \Omega_{ij}(1 - N_i) + N_k \Omega_{ik}(1 - N_i) - \Omega_{jk} N_j N_k$$

$$+ \Omega_{ijk} N_j N_k - 2N_i N_j N_k \Omega_{ijk} \quad , \tag{5}$$

where N_i is the atomic fraction of the ith element and Ω_{ijk} is a higher-order interaction parameter. The bonding energy of an alloy mixture can be written as the sum of the total energy resulting from each type of primary bond and that of the average energy influence on the primary bond. Thus, we can write Ω_{ij} in terms of the bond energies. Ω_{ij} is defined as

$$\frac{\Omega_{ij}}{z} = [E_{ij} - (E_{ii} + E_{jj})/2] \quad , \tag{6}$$

where z is the coordination number.

Pauling's Δ relation, expressing the difference between the actual bond energy and that predicted from additivity, is

$$\Delta = [E_{AB} - (E_{AA} + E_{BB})/2] = (\chi_A - \chi_B)^2 \quad , \tag{7}$$

where χ_A and χ_B are Pauling electronegativities.

6.2   The Interaction Strength Parameter, χ

Noting the similarity between Eqns. (6) and (7), we define in an analogous way an interaction strength parameter, χ expressible as

$$\chi^2 = (\varepsilon_j - \varepsilon_i)^2 = \frac{|\Omega_{ij}|}{n} \quad . \tag{8}$$

The constant n, which is the number of bonds in a molecule, may be derived from the formula for the first intermediate compound which

follows the solid solution region. For the case of complete solid
solubility, n is equal to unity. Examples illustrating the derivation of n from phase diagrams are presented below:

$$\text{Ti} + 1/3 \text{ Al} \overset{\rightarrow}{\leftarrow} 1/3 \text{ Ti}_3\text{Al} \quad (n = 3) \quad ,$$

$$1/2 \text{ Ti} + \text{Fe} \overset{\rightarrow}{\leftarrow} 1/2 \text{ TiFe}_2 \quad (n = 2) \quad ,$$

$$\text{Ti} + \text{Zr} \overset{\rightarrow}{\leftarrow} \text{Ti} + \underline{\text{Zr}} \quad (n = 1) \quad .$$

The electronegativity difference defined in this manner is given by

$$(\varepsilon_j - \varepsilon_i) = \left| \Omega_{ij}/n \right|^{1/2} = \chi_{ij} \quad . \tag{9}$$

The absolute value of Ω_{ij} is indicated since it can take on
both positive and negative values. The sign of Ω_{ij} itself is of
importance in the study of solid solution strengthening, as it
reflects the nature of the interatomic forces in alloy systems.
With $\Omega_{ij} < 0$ there is a tendency for SRO; and when $\Omega_{ij} > 0$, a
tendency for clustering exists.

After assigning a numerical value for the electronegativity of
Ti, it is possible to derive a set of electronegativity values for
various elements dissolved in it. By taking for the electronegativity of Ti the value $1.5\sqrt{\text{eV}}$ (which is its value on the Pauling
scale and also on a published [17] Soviet scale) the electronegativity scale of Table III was constructed.

The parameter which we have selected for subsequent use in this
work in establishing a relationship between the strengthening coefficient $(d\sigma/dc)$ and an electronic property will be derived from the
Table III electronegativity values. We refer to this parameter as
the "reduced interaction strength parameter," I. It is proportional
to the heat of mixing for the alloy, since the bonding enthalpy of
mixing [18] can be expressed as

$$\Delta H_{\text{bond}}^M = \Sigma \Delta H_{\text{bond}}^M (\text{binary systems}) + \alpha_{ijk} N_i N_j N_k \quad , \tag{10}$$

where α_{ijk} is a ternary interaction parameter. In the expression
for ΔH_{bond}^M, the first term, $\Sigma \Delta H_{\text{bond}}^M$(binary systems) is dominant.
The parameter, I, is proportional to this and is written as

$$I = \left| \pm I_{ij} \pm I_{ik} \pm I_{jk} \right| \quad , \tag{11}$$

where the respective I_{ij} have signed values based on whether the
ΔH^M for the particular binary system is positive or negative. In
case the thermodynamic data are not available for a particular

system, one can use the phase diagram to indicate the sign of ΔH^M
for the binary alloy [19]. It will be shown in a later section
that $d\sigma/dc$ is a function of I^2.

Table III. Electronegativity Scale for Ti-Base Systems

Element	ε_i($\sqrt{eV}$)	χ-Pauling($\sqrt{eV}$)	χ-USSR($\sqrt{eV}$)
Ti	1.50	1.50	1.50
Zr	1.60	1.40	1.60
Hf	1.61	1.30	1.61
Nb	1.61	1.60	
Cu	1.57	1.90	
Cr	1.70	1.60	
Ta		1.50	1.71
Mo	1.88	1.80	1.68
Fe	1.88		
V	1.78	1.60	1.83
Al	1.86	1.50	1.86
Ga	1.93	1.60	
Bi	1.95	1.90	
Sn	2.05	1.80	
O	3.88	3.50	3.88
N	4.50	3.00	

7. MECHANICAL PROPERTIES OF Ti-B-TYPE ALLOYS

Ti-B-type, or α-stabilized, Ti alloys are usually hcp up to
temperatures higher than the hcp $\rightleftarrows$ bcc ($\alpha \rightleftarrows \beta$) transformation temperature for pure Ti (viz. 882 - 885°C). The composition range of
single-phase-α solid solubility (in low-O-content, < 600 ppm, alloys)
is approximately 0-9 at.% Al [20]. Short-range ordering has been
observed within the composition range 9-12 at.% Al, corresponding
to the ($\alpha + \alpha_2$) phase boundary. The phase diagram [21] is presented
in Fig. 12. The reduced interaction strength, I, for the Ti-Al system is equal to 0.36$\sqrt{eV}$, which is consistent with the fact that the
solid solution region is terminated by a brittle intermetallic compound. Furthermore, by comparing the I-values for Ti-Ga and Ti-Sn,
we observe that I(Ti-Sn) > I(Ti-Ga) > I(Ti-Al), and their respective
values are 0.55$\sqrt{eV}$, 0.43$\sqrt{eV}$, and 0.36$\sqrt{eV}$. These three binary alloy
systems have comparable phase diagrams and have negative Ω_{ij} values.
The relative stabilities of the Ti_3B compounds where B stands for
Al, Ga, or Sn are shown in Fig. 13. Hume-Rothery [22] has observed
that electronegativity differences of about 0.3 are indicative of
fairly stable intermetallic compounds. This is confirmed by the

present results expressed in terms of the reduced interaction
strength parameter, I (Fig. 13).

The tensile strength, whether it be the proportional limit,
0.2% offset yield strength, or the ultimate tensile strength, turns
out to be proportional to $c^{1/2}$ (0 < c < 4 at.% solute). The data
for the Ti-Al, Ti-Ga and Ti-(Al_x-Ga_x) systems are presented in
Fig. 14. A number of investigators have studied yield strength-
concentration relationships in polycrystalline Ti-Al alloys. For
example, Ogden et al. [23] in an investigation on alloys containing 0-12.5 at.% Al, observed the room temperature yield strength
to increase with Al content, the resulting curve being concave
upwards. Blackburn and Williams [24] noted that the room temperature flow stress varied with Al concentration according to $c^{1.5-2}$
up to 10 at.% Al, and Truax and McMahon [25] observed similar
behavior in alloys of up to 13.5 at.% Al. These results are consistent with the existence of an expected SRO component to the

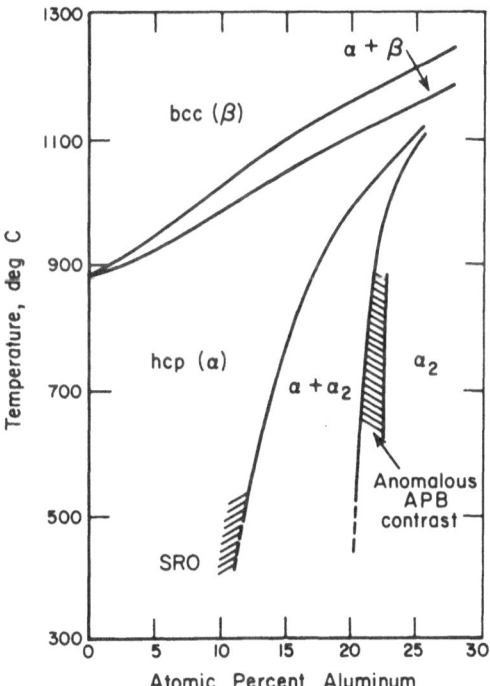

Fig. 12. Partial phase diagram for the Ti-Al system after
Blackburn [21]

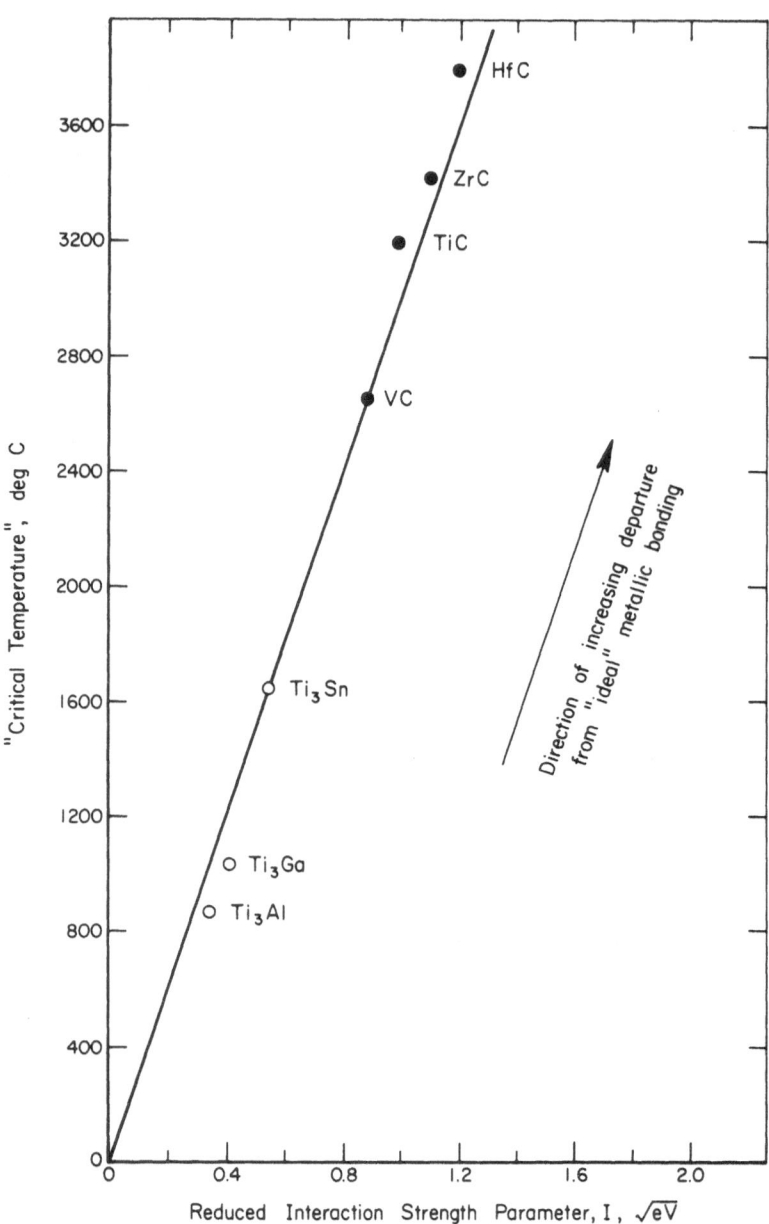

Fig. 13.   Critical temperatures, i.e., order-disorder or melting-point temperature, as a function of the reduced interaction strength parameter, indicating departures from "ideal" metallic bonding.

flow stress, in compositions greater than about 5 at.% Al, which should yield a $\sigma \propto c^2$ relationship. Evans [26] and Conrad [27-32] have investigated the deformation dynamics of Ti-Al monocrystalline and polycrystalline materials at low temperatures, and agree that the rate-controlling mechanism involves a Fleischer-type barrier.

Al strengthens α-Ti athermally without any significant effect on the thermally activated component of the flow stress. The athermal component is approximately 90% of the flow stress.

According to Conrad, the shear stress response to α-phase-stabilizing (e.g., Al, Ga, Sn) additions to Ti is given by

$$\tau = \tau^*(T,\gamma,c_i) + 0.5 \, \mu b \rho^{1/2} + \tau_\mu(c_s) \quad , \tag{12}$$

where τ is the resolved shear stress, T the absolute temperature, γ the shear strain rate, c_i the interstitial solute content, μ the shear modulus, b the Burgers vector, ρ the dislocation density, and c_s the substitutional solute content. For α-phase stabilizing

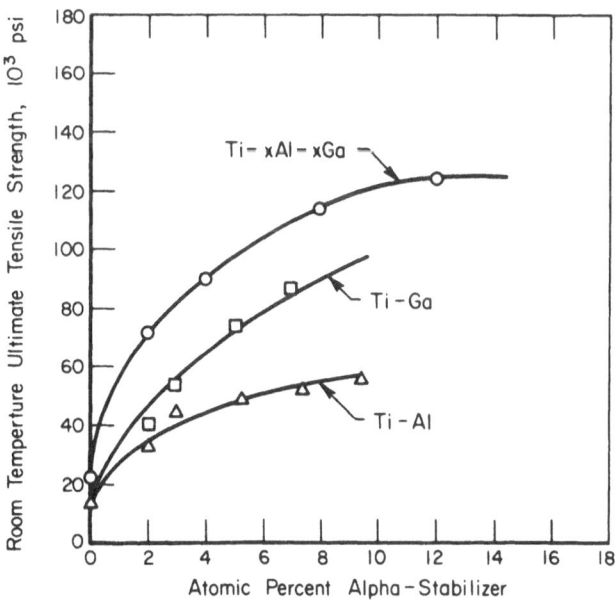

Fig. 14. Effect of ternary alloying on the room temperature strength of Ti alloyed with Al and Ga

solutes, the athermal shear stress is given by

$$\tau_\mu(c_s) = AIc_\alpha^{1/2} \ . \tag{13}$$

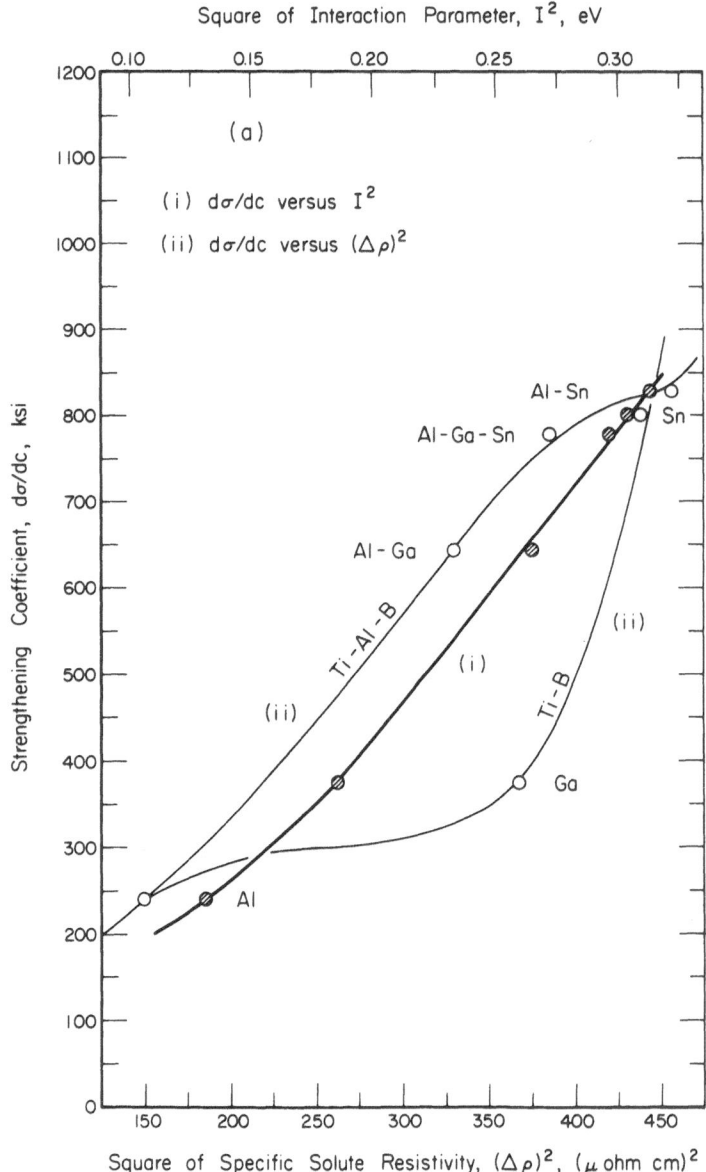

Fig. 15. Dependence of the strengthening coefficient, $d\sigma/dc$, on the reduced interaction strength parameter, I, for Ti-B alloys

Table IV. Rates of Strengthening, dσ/dc, Resistivity Increase, dρ/dc, and Values of the Thermodynamic Interaction Parameter, I, for B-Metals Dissolved in Ti

System Class	Alloy*	Strengthening Coefficient dσ/dc**, psi, or G/n†	Slope of ρ(c) Curve (μ ohm-cm)	Sq. of Specific Solute Resistivity $(\Delta\rho)^2$ $(\mu \text{ ohm-cm})^2$	Sq. of Interaction Parameter, I^2 (eV)
hcp (α)	Ti-Al	240 Gα/25	1220	149×10^4	0.13
	Ti-Ga	375 Gα/16	1918	368×10^4	0.19
	Ti-Al-Ga	642 Gα/9.4	1814	329×10^4	0.27
	Ti-Al-Ga-Sn	779 Gα/7.7	1961	385×10^4	0.30
	Ti-Sn	800 Gα/7.5	2093	438×10^4	0.30
	Ti-Al-Sn	825 Gα/7.3	2138	457×10^4	0.31
	Ti-Ge	5000 Gα/1.2	2254	508×10^4	0.72

* B-metal solutes-equiatomic

** Δρ = resistivity changes per at.% total B-metal solute

† Shear-modulus notation; Gα = 6000 psi

Strengthening coefficients for various α-phase stabilizing elements are presented in Table IV, and summarized in Fig. 15 which shows the $d\sigma/dc$ <u>versus</u> I^2 and $d\sigma/dc$ <u>versus</u> $(\Delta\rho)^2$ relationships. These results demonstrate that B-metals are moderately rapid strengtheners of Ti in accordance with Fleischer's [33] rapid hardening model. The curves also illustrate the dependence of the strengthening coefficient, $d\sigma/dc$ on (a) the reduced interaction parameter, I, and (b) the specific resistivity, $\Delta\rho$, for B-elements in Ti. The rapid strengthening, we assert, can be traced to a relatively large atomic potential difference between the solute and solvent ions (i.e., the $E_{1,2}$ referred to earlier. Both $(\Delta\rho)^2$ and I^2 are measures of this difference. Making a connection with alloy theory, we point out that rapid strengthening in response to the addition of a given solute species is associated with pronounced departures from "rigid-band" behavior. Since the athermal component of the shear stress is at least 90% of the shear stress, it is reasonable to state that a $\tau_\mu(c_s) \sim c^{1/2}$ law is obeyed for all of the concentrations investigated.

8. STRENGTHENING MECHANISMS

As indicated in the earlier sections, a relatively large $E_{1,2}$ leads to a significantly structure-dependent free energy. Microscopically this can be ascribed to directional bonding such as that associated with a dissolved B-metal in Ti. A possible strengthening mechanism involves a dislocation-pinning interaction which will occur during the time that several atoms along the line of a moving dislocation become a part of some directionally bonded local group. This electronically derived mechanism can be identified with Fleischer's [33] "rapid hardening" situation. It is, moreover, interesting to note that Fleischer required impurity atoms to occupy sites of reduced symmetry, but did not specify whether they need be interstitial or substitutional. Our electronic model is entirely consistent with this concept, directional bonding associated with the presence of the dissolved impurity furnishing a low-symmetry environment. In this electronic model, no essential distinction need be drawn between interstitial and substitutional site occupancy.

In Ti-base alloys, for example B-element (i.e., s, p element) solutes, whether substitutional or interstitial, induce a bulk stiffening of the Ti matrix; together with "rapid" strengthening, brought about <u>via</u> the dislocation-pinning mechanism, referred to above, arising from directional electronic bonding associated with the presence of the B-element impurity. The bonding properties of these local atomic groupings can be deduced to a crude approximation by a consideration of the electronic properties and structures of the first intermetallic compound. The electronic and physical

properties of this compound are felt even in the dilute region of
the solid solution phase. Based on the structure and relative
phase stability of the first intermetallic compound, one can make
qualitative predictions, from the phase diagram, about the behavior
of, for example, the elastic constants during alloying.

Alloy systems such as Fe-Co, Fe-Cr, and Fe-Re, shown in Fig. 16
(a,b,c), all stabilize the room temperature α-phase of Fe (bcc in
this case) to reasonably high temperatures, and the solid solution
region is terminated by a compound having reasonable stability.
The Fe-Co system has the α-solid solution phase stabilized to
∿985°C, and this region is terminated by the Fe-Co ordered compound
with an Fe-Co → α transformation temperature of 730°C. The Fe-Cr
system is characterized by a γ-loop; and a solid solution α-region
terminated by σ-phase, a hard, brittle compound, with a σ → α transformation temperature of ∿815°C. The Fe-Re system is comparable to
the Ti-Al, Ti-Ga, and Ti-Sn alloys in that it is characterized by a
peritectoid phase transformation at 895°C, i.e., γ + ε → α. The
ε-phase is Re_2Fe_3 which is stable to ∿1500°C. Thus one would predict that the Young's modulus, E, and the shear modulus, G, should
both increase in the sequence Co < Cr < Re, which is indeed found
to be the case [34,35]. Leslie [35] noted that the rule of mixtures does not seem to be obeyed here since it is possible for
elements with greater moduli than Fe, when dissolved in it, to
yield alloys with lower moduli than the host. The failure of the
rule of mixtures is probably related to a lattice stability situation which may compete with solid solution strengthening. The
lattice stability of any cubic solid solution phase can be judged
from the behavior of the elastic shear modulus $C' = (C_{11} - C_{12})/2$,
and the manner in which it changes with alloying (or electron/atom
ratio, ζ), as indicated in Fig. 5. Lütgering and Hornbogen [36]
also observed that the hardening of Fe by Co and Cr followed the
Fleischer model, behavior which is consistent with the electronic
approach to solid solution strengthening.

The work by Dey, Gilman, and Nehrenberg [37] on untempered
Fe-Ni-C martensites also supports the concept of a localized-
directional-bond model of solid solution hardening. They observed
that both E and G for the untempered martensites show marked increases (∿33%) following the addition of C to the lattice. This
important result suggests that the intrinsic cohesive forces within
martensite crystals increase with C content. Hardness depends on
local cohesion maxima and the stiffness of the C-lattice interaction. This increase in local stiffness was estimated by extrapolat-
ing the alloy shear modulus to 100% C. The resulting shear modulus
was only about 30% less than that of diamond.

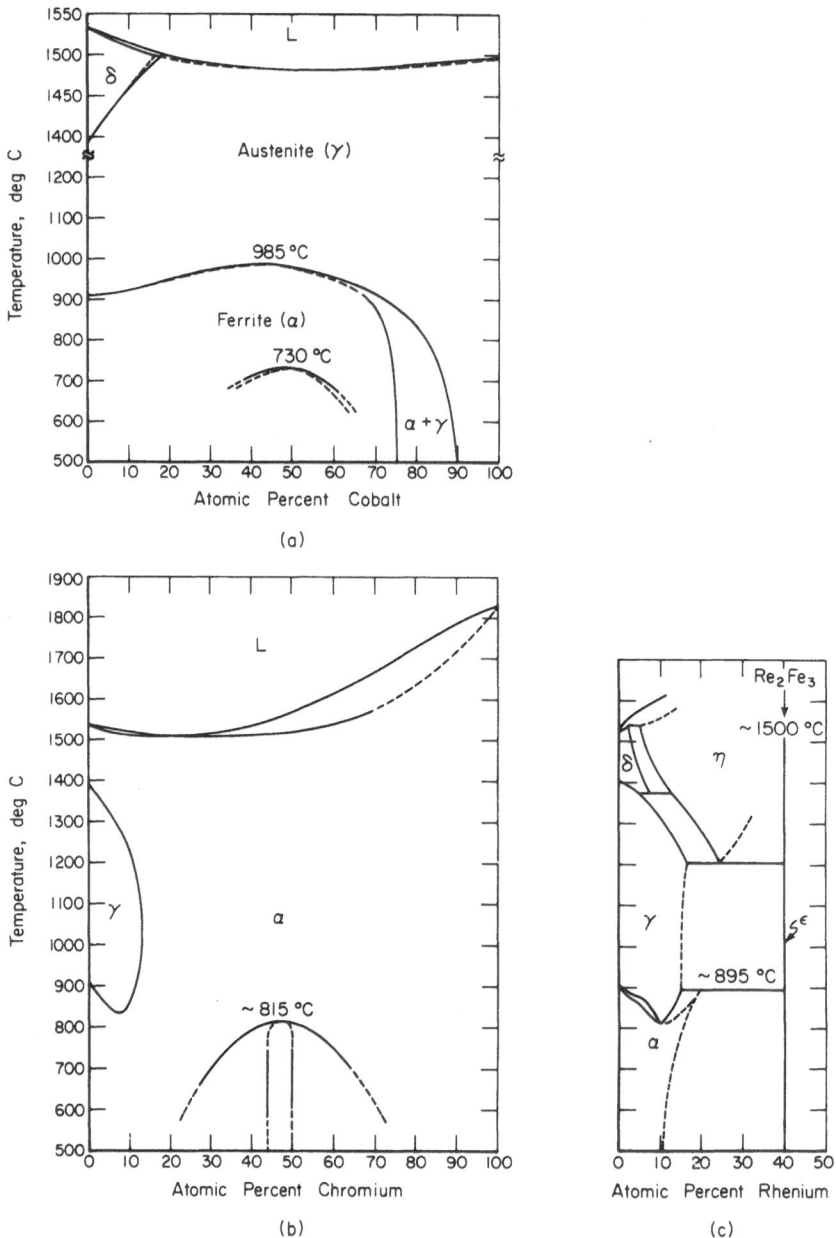

Fig. 16. Phase diagrams for Fe-Co, Fe-Cr, and Fe-Re alloy systems

9. MECHANICAL BEHAVIOR OF Ti-T$_2$-TYPE ALLOYS AND OTHER BCC TRANSITION METAL ALLOYS

The mechanism of solid solution hardening in bcc Ti-base alloys is complicated by the problem of lattice stability. Stabilization of the β-phase of titanium can be achieved by increasing the average electron/atom ratio through alloying with a "nearby" transition element. In general, the perturbation of electron states accompanying the mixing of transition elements is rather small particularly for adjacent transition elements near the middle of a period [38,39], and this perturbation does not lead to significant solution strengthening. Size-effect strengthening and lattice-stability considerations tend to dominate. As an example of binary transition-metal alloying, McMillan [40] has shown that binary alloys from the sequence Hf-Ta-W-Re exhibit excellent rigid-band behavior. In Ti-T$_2$ alloys the pairwise thermodynamic interaction parameter, Ω_{ij}, is always positive; and the $\varepsilon_i - \varepsilon_j$ values are relatively small, e.g., $\varepsilon_{Zr} - \varepsilon_{Ti} = 0.1$ and $\varepsilon_V - \varepsilon_{Ti} = 0.28$. In contrast to the Ti-B which are SRO systems, Ti-T$_2$ alloys exhibit a tendency for clustering.

Lattice stability can be studied using various model systems such as Ti-Mo or Ti-V. Instability of the bcc structure is expressed in terms of the establishment of a 2/3 <111> longitudinal displacement wave [41,42,43], a condition which leads to recognizable ω-phase precipitation in alloys for which the electron/atom ratios ($\mathfrak{z}$) are suitable, e.g., 4.1 to 4.3 in the case of Ti-Mo. Absolute instability of the bcc structure accompanies the vanishing of the elastic shear modulus $C' = (C_{11} - C_{12})/2$ which occurs generally for $\mathfrak{z} \gtrsim 4.1$ (Fig. 5).

The problem of solution strengthening, as well as alloy toughness, in transition-metal binary alloys is intimately related to the stability of the bcc structure. In order to study solid-solution hardening of β-Ti alloys, it was first necessary to achieve a stable bcc lattice. This was accomplished by preparing a Ti-Mo (25 at.%) master alloy, i.e., an alloy whose composition falls outside the range where the lattice is unstable or where ω-phase forms. The solutes Ge, Sn, Al, and Ga were added respectively to the master alloy for the purpose of determining the strengthening coefficients, $d\sigma/dc$. These solute elements in the pseudo-β alloy did not follow a $c^{1/2}$ law as they did in the dilute α-Ti alloys; rather, they followed a linear relation. Since Ti-Mo alloys are prone to clustering, a contribution to strengthening from this effect may mask the $c^{1/2}$ behavior of the nontransition elements in the dilute region.

The relationship between the strengthening coefficient, $d\sigma/dc$, and the I^2 and $(\Delta\rho)^2$ parameters for these alloys is presented in Fig. 17 and the data are summarized in Table V. These results

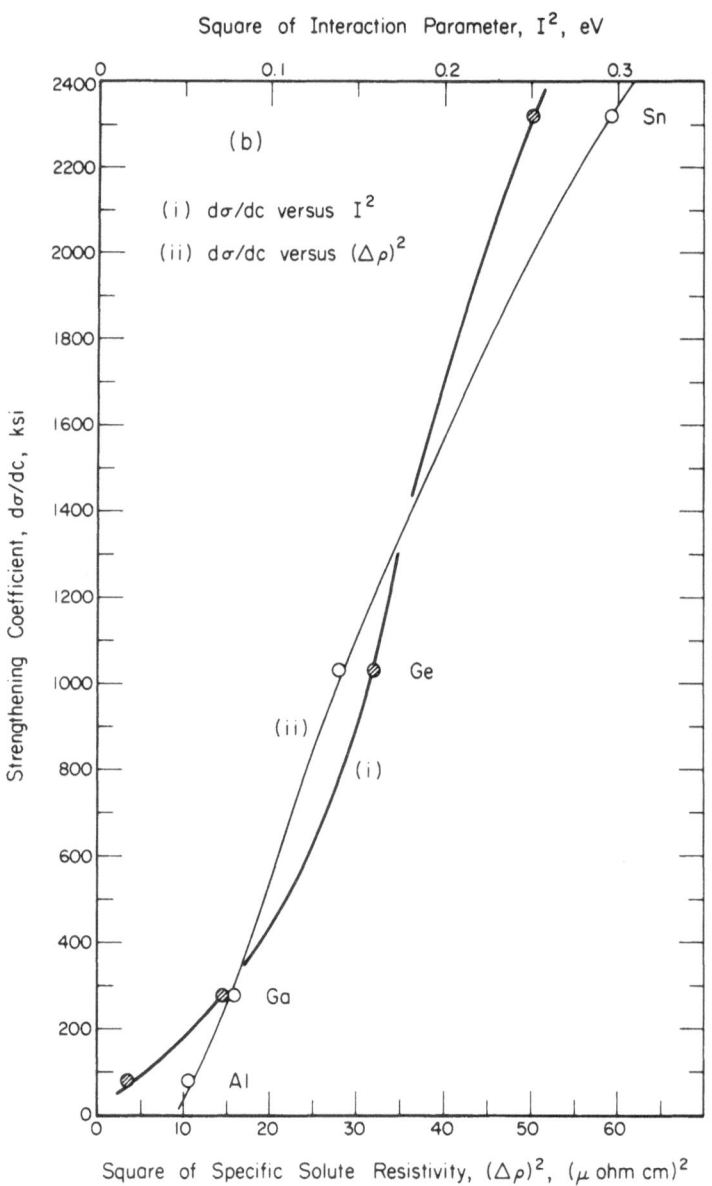

Fig. 17. Dependence of the strengthening coefficient, dσ/dc, on the reduced interaction strength parameter, I, for Ti-Mo(25 at.%)-B alloys

Table V. Rates of Strengthening, $d\sigma/dc$, and Resistivity Increase, $d\rho/dc$, and Values of the Thermodynamic Interaction Parameter, I, for B-Metals Dissolved in Ti-Mo (25 at.%)

System Class	Alloy*	Strengthening Coefficient $d\sigma/dc$**, ksi, or G/n†	Slope of $\rho(c)$ Curve (μ ohm-cm)	Sq. of Specific Solute Resistivity†† $(\Delta\rho)^2$ (μ ohm-cm)^2	Sq. of Interaction Parameter, I^2 (eV)
bcc (β)	(Ti-Mo)-Al	78 Gβ/40	330	11	0.02
	(Ti-Mo)-Ga	275 Gβ/13.5	400	16	0.07
	(Ti-Mo)-Ge	1025 Gβ/3.6	529	28	0.16
	(Ti-Mo)-Sn	2320 Gβ/1.6	770	59	0.25

* B-metal-equiatomic; beta-alloy base, 25 at.% Mo
** c represents mole-fraction of solute (<7 at.%)
† Shear-modulus notation; Gα = 6000 ksi, Gβ = 3800 ksi
†† $\Delta\rho$ = resistivity changes per at.% total B-metal solute

confirm the postulate that B-metals, which are moderately rapid
strengtheners of Ti, should perform with comparable effectiveness
in a bcc Ti-base alloy. In shear-modulus representation, hcp and
bcc alloys experience comparable B-element strengthening, the rates
of which vary between about G/1.5 and G/n > 10 in both classes of
system. In terms of $d\sigma/dc$ itself, the range of values covered is
comparable, excepting that Ge is an unusually potent strengthener
of hcp Ti, while Al performs rather poorly in Ti-Mo (25 at.%).
With the same mutually consistent exceptions, the ranges of I-values
covered are also comparable in both hcp and bcc system classes. In
conformity with earlier results for binary Ti-base alloys, the
appropriate columns of Tables IV and V and the $d\sigma/dc$ versus I^2
curves of Figs. 15 and 17, which increase monotonically from left
to right, demonstrate that I is also a valid indicator of solid
solution strengthening in both multicomponent hcp alloys and in bcc
Ti-Mo (25 at.%)-base alloys. The mechanical behavior of the Ti-Mo
(25 at.%)-base alloys support the results of low-temperature calori-
metry experiments on ternary alloys based on Ti-Mo (15 at.%). Those
results presented in Fig. 18 showed that when Al was added, Θ_D
increased while γ and T_c were decreasing. When Fe was substituted
for Mo, Θ_D increased with corresponding decreases in γ and T_c.
This behavior was predictable from the position of Fe in the transition metal series. Al both strengthens and stabilizes the bcc
lattice; the superconducting transition temperature decreases, and
at the same time, the tendency for ω-phase precipitation is inhib-
ited. In conclusion, we take note of some of the numerous published
studies of strengthening in bcc transition metal alloys. For example, the following metals and alloys have been investigated: Cr [44],
Ni and Pt in Fe [45]; Re in W [46] and Mo [47], Nb [48], W, Mo [49]
and Re in Ta [50]; Ta, V, Zr and W in Nb [51]; and Hf, Ta, W, Re,
Os, Ir and Pd in Mo [52].

Speich, Schwoeble, and Leslie [34] have determined the isotropic elastic constants of Fe alloyed with C, Mn, Ni, Cr, Co, Ir,
Pt, Re, Rh, and Ru for the purpose of understanding solid solution
strengthening. They found that $\tau_\mu(c_s)$ tended to follow a c or c^2
relationship in general, but there is some evidence favoring
$\tau_\mu(c_s) \propto c^{1/2}$ at small concentrations. This general behavior is in
agreement with the $\tau_\mu(c_s) \propto c$ behavior for the B-metal solutes in
Ti-Mo (25 at.%) alloys.

Harris [51] has pointed out that isoelectronic alloys of Nb,
V, and Ta have size misfit parameters as the dominant term in solid
solution strengthening. When other solute elements are added to
this isoelectronic series, the stability of the lattice is further
perturbed. Based upon the position of Nb, V, and Ta on the C' versus
$\mathfrak{z}$ curve, the addition of Mo, which increases $\mathfrak{z}$, causes the stability
of the lattice to increase in the direction of peak lattice sta-
bility; i.e., towards Cr, Mo, and W. This statement is in agreement

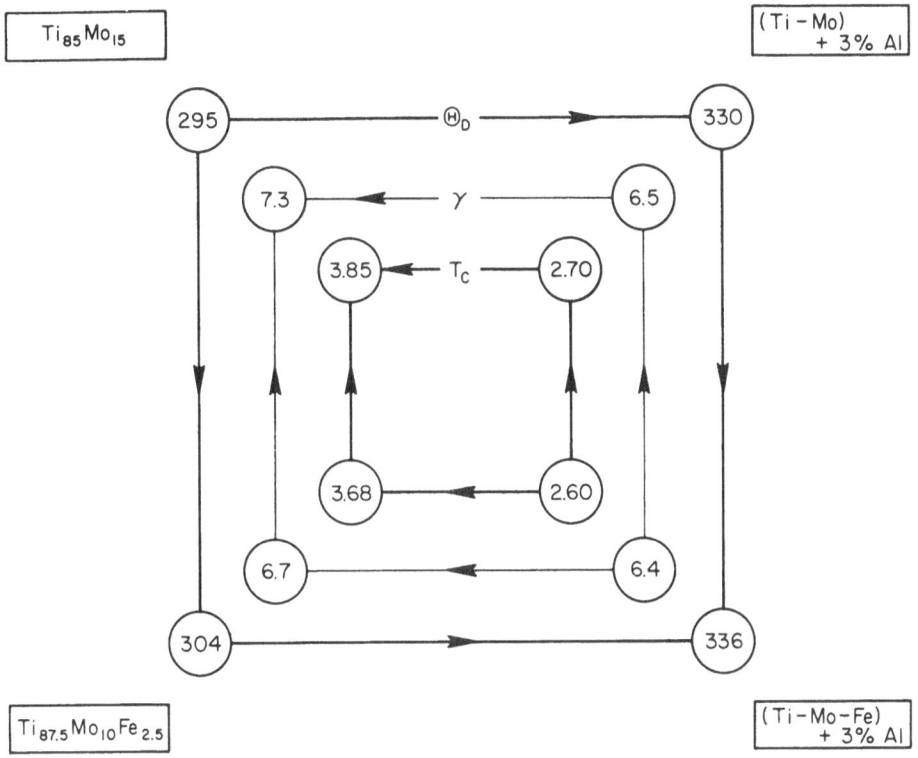

Fig. 18. The influence of small additions of Al and Fe on some representative mechanical (Θ_D) and electronic (γ, T_c) properties of Ti-Mo alloys at constant electron-to-atom ratio

with the observation of straight dislocations in the Ta-Mo system by Van Torne and Thomas [49]. On the other hand, these same alloys when alloyed with Ti, for example, go in a direction of increasing lattice <u>instability</u>, a process which leads to, and is interrupted by ω-phase precipitation. Likewise, the alloying of W, Mo, and Cr with Re causes the bcc lattice to become unstable and consequently more ductile, especially in the neighborhood of $\gamma \sim 6.3$. Eventually, this lattice instability, which is responsible for the ductilization of these Group VI elements, is interrupted by the formation of σ-phase for $6.4 \lesssim \quad \lesssim 6.8$. The σ-phase is very brittle.

We can summarize the observations on strength and stability
of the bcc lattice by stating that such a lattice has an instability towards directional bonding. Strengthening in alloy
systems can be judged qualitatively on the basis of their binary
phase diagrams and the phase stability of the compounds which form.
It is also for these reasons that the reduced thermodynamic interaction strength, I, is a valid parameter descriptive of the behavior
of the strengthening coefficient, $d\sigma/dc$. This parameter has both
phase and lattice stability terms built into it, since it is proportional to ΔH^M_{bond} [52].

ACKNOWLEDGMENTS

The research of one of the authors (H.L. Gegel) was conducted
under the auspices of the Air Force Materials Laboratory, Wright-
Patterson Air Force Base, Ohio; and that of the other (E.W. Collings)
was supported by the Air Force Office of Scientific Research (Air
Force Scientific Command) under Grant No. OSR-71-2084.

REFERENCES

1. COHEN, M.H., in Alloying Behavior and Effects in Concentrated
    Solid Solutions, MASSALSKI, T.B., ed., Gordon and Breach, p 1,
    1965.
2. COLLINGS, E.W., ENDERBY, J.E. and HO, J.C., in Proc. 3rd
    Materials Research Symposium, "Electronic Density of States",
    Nat. Bur. Stand. (U.S.) Spec. Publ. 323, p 483, December 1971.
3. COLLINGS, E.W., ENDERBY, J.E., GEGEL, H.L. and HO, J.C., in
    Titanium Science and Technology, Vol. 2, JAFFEE, R.I. and
    BURTE, H.M., eds., Plenum Publishing Corp., New York, p 801,
    1973.
4. GEGEL, H.L., HO, J.C. and COLLINGS, E.W., "An Electronic
    Approach to Solid Solution Strengthening in Titanium Alloys",
    in Proc. 3rd International Conference on the Strength of Metals
    and Alloys, sponsored by the Inst. of Metals, ASM, and Japan
    Inst. of Metals, Cambridge, England, August 1973.
5. COLLINGS, E.W. and GEGEL, H.L., Scripta Met. 7 437 (1973).
6. COHEN, M.H., in Colloquium on Solid Metallic Solutions, Orsay,
    1972; J. Phys. Radium, 23 643 (1962).
7. BLANDIN, A., in Alloying Behavior and Effects in Concentrated
    Solid Solutions, MASSALSKI, T.B., ed., Gordon and Breach, p 50,
    1965; PICK, R. and BLANDIN, A., Phys. kondens. Materie, 3 1
    (1964).
8. HARRISON, W.A., Pseudopotentials in the Theory of Metals,
    Benjamin, Inc., New York, 1966.

9.   GEHLEN, P.C., in The Science, Technology and Application of
     Titanium, JAFFEE, R.I. and PROMISEL, N.E., eds., Pergamon
     Press, New York, p 349, 1970; HO, J.C., GEHLEN, P.C. and
     COLLINGS, E.W., Solid State Comm., Vol. 7, p 511, 1969.
10. FOWLER, R. and GUGGENHEIM, E.A., Statistical Thermodynamics,
     Cambridge University Press, Cambridge, 1939.
11. GEGEL, H.L. and HOCH, M., in Titanium Science and Technology,
     Vol. 2, JAFFEE, R.I. and BURTE, H.M., Plenum Publishing Corp.,
     New York, p 923, 1973.
12. ROLINSKI, E.J., HOCH, M. and OBLINGER, C.J., Met. Trans. 3
     1413 (1972).
13. ROLINSKI, E.J., HOCH, M. and OBLINGER, C.J., Met. Trans. 2
     2613 (1971).
14. HACKWORTH, J.V., HOCH, M. and GEGEL, H.L., Met. Trans. 2 1799
     (1971).
15. HOCH, M. and USELL, R.J., Jr., Met. Trans. 2 2667 (1971).
16. ROLINSKI, E.J., OBLINGER, C.J. and HOCH, M., Advances in Mass
     Spectrometry, Vol. 5, The Inst. of Petroleum, Adlard and Son
     Ltd., Dorking (Great Britain), p 408, 1971.
17. ASTAKHOV, K.V., Collection "Electronegativity", Western
     Siberian Publishing House, p 5, 1965, (as quoted in) GLAZUNOV,
     S.G., "Present Day Titanium Alloys", Titanium Alloys for New
     Technology, BAYKOUA, A.A., ed., U.S. Army Science and Technology
     Center Translation (FSTC-HT-23-581-69), p 7.
18. SHARKEY, R.L., POOL, M.J. and HOCH, M., Met. Trans. 2 3039
     (1971).
19. KUBASCHEWSKI, O., in Phase Stability in Metals and Alloys,
     RUDMAN, P.S., STRINGER, J. and JAFFEE, R.I., eds., McGraw-Hill
     Book Co., New York, p 63, 1967.
20. SAKAI, T., Plastic Deformation of Ti and Ti-Al Alloy Single
     Crystals, Ph.D. Dissertation, Northwestern University, Evanston,
     Illinois, June 1973.
21. BLACKBURN, M.J., Trans. AIME 239 1200 (1967).
22. HUME-ROTHERY, W., in Phase Stability in Metals and Alloys,
     RUDMAN, P.S., STRINGER, J. and JAFFEE, R.I., eds., McGraw-Hill
     Book Co., New York, p 3, 1967.
23. OGDEN, H.R., MAYKUTH, D.J., FINLAY, W.L. and JAFFEE, R.I., J.
     Metals, 267 (1953).
24. BLACKBURN, M.J. and WILLIAMS, J.C., Trans. Quart. 62 398 (1969).
25. TRUAX, D. and McMAHON, C., Plastic Properties and Fracture of
     Ti-Al Alloys, Final Rpt. N00019-69-C-0252, University of Pennsyl-
     vania, August 1970.
26. EVANS, K.R., Trans. AIME 242 648 (1968).
27. CONRAD, H., Acta Met. 14 1631 (1966).
28. CONRAD, H., Can. J. Phys. 45 581 (1967).
29. CONRAD, H. and JONES, R., The Science, Technology, and
     Application of Titanium, JAFFEE, R.I. and PROMISEL, N.E., eds.,
     Pergamon Press, New York, p 489, 1970.
30. KRATOCHVIL, J. and CONRAD, H., Scripta Met. 4 815 (1970).
31. TANAKA, T. and CONRAD, H., Acta Met. 20 1019 (1972).

32. OKAZAKI, K. and CONRAD, H., Trans. JIM 13 205 (1972).
33. FLEISCHER, R.L. and HIBBARD, W.R., The Relation Between the
     Structure and Mechanical Properties of Metals, H.M. Stationery
     Office, p 262, 1963.
34. SPEICH, G.R., SCHWOEBLE, A.J. and LESLIE, W.C., Met. Trans. 3
     2031 (1972).
35. LESLIE, W.C., Met. Trans. 3 5 (1972).
36. LÜTGERING, G. and HORNBOGEN, E., Z. Metallk. 59 29 (1968).
37. DEY, B.N., GILMAN, J.J. and NEHRENBERG, E.A., Phil. Mag. 24
     1257 (1971).
38. STERN, E.A., Energy Bands in Metals and Alloys, BENNETT, L.H.
     and WABER, J.T., eds., Gordon and Breach, p 151, 1968.
39. FRIEDEL, J., discussion in Phase Stability in Metals and Alloys,
     RUDMAN, P.S., STRINGER, J. and JAFFEE, R.I., eds., McGraw-Hill
     Book Co., New York, p 162, 1967.
40. McMILLAN, W.L., Phys. Rev. 167 331 (1968).
41. deFONTAINE, D., Acta Met. 18 275 (1970).
42. deFONTAINE, D., PATON, N.E. and WILLIAMS, J.C., Acta Met. 19
     1153 (1971).
43. SASS, S.L., Acta Met. 17 813 (1969).
44. HORN, G.T., ROY, P.B. and PAXTON, H.W., J. Iron Steel Inst.
     201 161 (1963).
45. KRANZLEIN, H.H., BURTON, M.S. and SMITH, G.V., Trans. AIME 233
     64 (1965).
46. GARFINKLE, M., NASA Techn. Rpt. X-52132 (1965).
47. LAWLEY, A. and MADDIN, R., Trans. AIME 224 573 (1962).
48. RUDOLPH, G. and MORDIKE, B.L., Z. Metallk. 58 708 (1967).
49. Van TORNE, L.I. and THOMAS, G., Acta Met. 14 621 (1966).
50. MITCHELL, T.E. and RAFFO, P.L., Canad. J. Phys. 45 1047 (1967).
51. HARRIS, B., phys. stat. sol. 18 715 (1966).
52. HOCH, M., private communication, September 28, 1973.

APPLICATION OF ALLOY PHYSICS TO SOLUTION STRENGTHENING

E. A. Stern

Department of Physics
University of Washington
Seattle, Washington

ABSTRACT

Experiments indicate that in Ti alloys, solution hardening
occurs predominantly by an electronic mechanism, i.e., by interactions within atomic dimensions of the impurity. The basic cause
of such a mechanism is investigated using ideas originating from
alloy physics. It is shown that the solution strengthening of
group B metals in Ti is dominated by the behavior of the d states
of the Ti host. They are repelled in the vicinity of the impurities, causing a large change in residual resistance which is experimentally correlated with good solution hardening properties. The
discussion shows that, in general, better solution hardening should
occur the further apart the columns of the solute and solvent are in
the periodic table, i.e., the greater their valence difference is.
A simple model calculation is made of the pinning capability of an
impurity under dilation strains, and qualitative agreement with
experiment is found.

1. INTRODUCTION

The strengthening of materials is of obvious practical importance and many different mechanisms for such strengthening have been
distinguished empirically. Of these mechanisms, the simplest in
terms of its relationship to fundamental processes is solution
strengthening or hardening. In solution hardening, impurities which
alloy with the host serve to strengthen it.

In recent years significant advances in our understanding of
alloy theory have occurred and it appears profitable to try to apply

some of these ideas to understand both qualitatively and quantitatively the basic mechanisms that produce solution hardening. This paper consists of a modest attempt along these lines with the goal of pointing out some of the necessary calculations which would have to be made in the future in order to quantitatively calculate solution hardening effects. The emphasis in this paper will be in trying to isolate the most important physical mechanisms contributing to solution hardening using basic physical ideas developed from alloy theory.

In a general way the basic idea of solution hardening was first introduced by Cottrell [1] who pointed out that impurities will have a different solubility in the vicinity of dislocations than between them. The solubility distribution will be such as to minimize the energy of the alloy. As dislocations move under external stress the minimum energy configuration of the impurities relative to the dislocation is disturbed, raising the energy, and, thus, acting as a pinning mechanism for such dislocation motion.

This picture is generally accepted but the actual detailed mechanism of the interaction of impurities with dislocations has never been completely isolated. Several candidates have been proposed but no general agreement exists for this interaction mechanism. It has been suggested that the interaction mechanism is either elastic in origin or electronic in origin [2]. Of course, basically even elastic energy can be traced to an electronic origin but by elastic will be meant that which can be described by classical elastic theory, namely a long-range interaction; while by electronic we will mean that which is short range in interaction, of the order of interatomic distances. One elastic candidate is the interaction between the elastic strain field introduced by the impurity atom, which has a different size than the host, and the elastic strain field surrounding dislocations. Such a mechanism would predict that strengthening depends only on the volume change per atom introduced by the impurity. In the electronic category is placed the modulus defect mechanism where impurities change the elastic constants of the material. This modulus defect can interact with either the core of the dislocation or its long range elastic field.

Recent experiments by Collings and Gegel [3] are very important in that they narrow the range of acceptable mechanisms for solution strengthening of Ti-based alloys. These experiments have shown a correlation between the interaction parameter Ω_{ij} (a measure of the interaction between impurity and host) and strengthening. In particular, these experiments give a very strong indication that the mechanism for solution strengthening in Ti-based alloys is electronic and short ranged, i.e., of the order of interatomic dimensions.

This follows from the fact that both residual resistance and Ω_{ij} are electronic in nature and dominated by effects localized within interatomic dimensions. Electrical resistance has its major contributions from those components of the perturbing potential which vary appreciably within dimensions of the same order or smaller than the Fermi wavelength of the scattered electrons. For metals, the Fermi wavelength is of atomic dimensions. In agreement with this, Ω_{ij} is dominated by nearest neighbor interactions.

Based on this experimental information plus the general ideas of Cottrell [1] we propose the following model for solution strengthening of Ti alloys, and we will proceed to check its feasibility in the rest of the paper. Depending on their detailed interactions, impurities preferentially dissolve either in the dislocation core or away from it. As the dislocation moves, or attempts to move, the environment surrounding a given impurity will change. Such a change will disturb the minimum energy configuration and raise the energy of the system pinning the dislocation. The strengthening will be greater, the larger the increase of energy caused by a given distortion around the impurity. Since the experiments indicate that the concerned interaction is short ranged, the interaction with nearest neighbors should be dominant. We make the simplifying assumption of neglecting entirely the effects of neighbors farther from the impurity than the 1st nearest neighbors.

With this assumption the question of solution hardening reduces to calculating the change in energy of an impurity as its nearest neighbors vary their relative positions. From our assumption of neglecting the effect of second, third, etc., nearest neighbors, we have the freedom of choosing their configuration to simplify the problem. We do so by arranging them so that the distortion around the impurity is part of a macroscopic and uniform distortion of the surrounding host.

In other words the distortion is produced by first considering a single impurity in the host, undistorted except for any elastic distortion produced by a size mismatch. Then a macroscopic and uniform distortion ΔS is applied to the host so as to produce the desired distortion around the impurity. Changes in energy of the system versus distortion will be calculated and ΔE_i, that part due to the presence of the impurity, will be isolated. Such a model neglects the interaction between impurities and is thus valid only for dilute concentrations. This is not a serious restriction since the dilute range is that range where the strengthening varies linearly with solute concentration and will characterize the hardening property of the solute. The magnitude of $\Delta E_i / \Delta S$ is a measure of the solution strengthening, the larger $\Delta E_i / \Delta S$ is, the greater the solution strengthening.

Following this, Section 2 introduces some basic and simple con-
cepts of alloy theory and uses them to estimate the change in residual
resistance introduced by the impurities, obtaining agreement with
experimental values. These estimates show that the d electrons are
dominant in producing solution hardening in transition metal hosts.
Section 3 contains a simple model calculation of the pinning energy
of an impurity against volume change distortions. A discussion and
summary is presented in Section 4.

2. ALLOY THEORY CONCEPTS

The experiments of Collings and Gegel [3] show that solution
strengthening increases roughly in proportion to the distance between
the columns in the periodic table of the solvent and solute. In
trying to understand this result theoretically, the question naturally
arises as to what is the state of the alloy around such an impurity.
We consider, to be specific, systems studied by Gegel and Collings,
namely Ti as the host and Ga, Al, and Sn as the impurities. Table I
gives the atomic configuration of the valence electrons for these
elements.

Table I. Atomic Configurations

$$Ti = 3d^2 \, 4s^2$$
$$Ga = 4s^2 \, 4p$$
$$Al = 3s^2 \, 3p$$
$$Sn = 5s^2 \, 5p^2$$

One striking difference between the host and the impurities is
the lack of d-character in the impurity valence states. When an
impurity is placed in Ti, its lack of d-character causes a large
perturbation in the wave function of the system in its vicinity. To
understand this we make some simplifying assumptions which will re-
veal the physical basis of the effect. Although these simplifying
assumptions introduce some approximations, they do not change the
qualitative behavior which we are most interested in exposing at
this time. The assumptions we make are: (1) The hybridization [4]
between d and s-p electron states is neglected. We do include the
hybridization between s and p states by assuming that the s and p
states are completely mixed so that there are no separate s and p
bands but a single s-p band. The neglect of hybridization between
d and s-p electrons produces about a 20% quantitative error since
for the transition metals the hybridization interaction is about
20% of the d-band width. (2) The configuration change in going
from the atom to the solid is neglected, i.e., we assume that in the
solid the character of the wave function in the vicinity of an atom

is as given in Table I, the atomic configuration. This assumption
is incorrect as it is generally believed that in the solid a transition metal has about one s electron instead of two [5] as in atomic
Ti, and thus greater d-character. However, since we are interested
in understanding the basic physics and only estimating the magnitude
of effects, such an assumption will be sufficient, though it is more
correct to use the electronic configuration in the solid. (3) Charge
neutrality within a Wigner-Seitz cell surrounding the impurity is
assumed. The Wigner-Seitz cell is the smallest one formed by bisect-
ing with planes the lines between the impurity and its nearest neighbors. Such cell is indicated in Fig. 1. Expected deviations from
charge neutrality in metals are of the order of tenths of an electronic charge, again not too serious an error. (4) The final assumption is a sort of muffin-tin approximation [6] where the perturbation
introduced by the impurity is assumed to be spherically symmetric,
highly localized (already assumed in (3)), and between the impurity
and its nearest neighbors there is a region where the potential is
approximately constant. This then permits the defining of phase
shifts introduced by the presence of the impurity [7]. Again such
an assumption is rough, but sufficient for our qualitative, semi-
quantitative discussion.

     From the assumption (1) we can represent the Bloch wave func-
tion $\psi_{k\ell}$ of the pure host using a Wannier representation as

$$\psi_{k\ell} = \sum_n e^{i\vec{k}\cdot\vec{R}_n} \phi_\ell \; (\vec{r}-\vec{R}_n), \qquad (1)$$

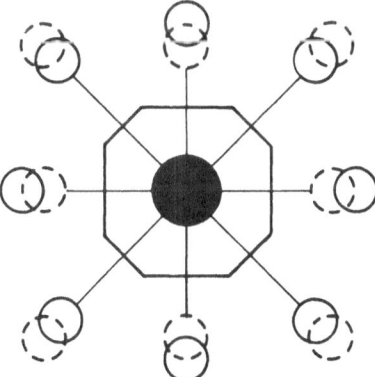

Fig. 1. Illustrating a Wigner-Seitz cell around an impurity. The
dotted circles represent the positions of the atoms before the introduction of the impurity at the center. The solid circles represent
the location of the surrounding atoms after introduction of the
impurity.

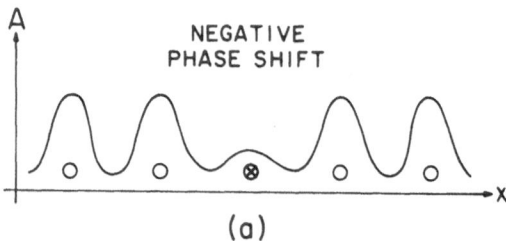

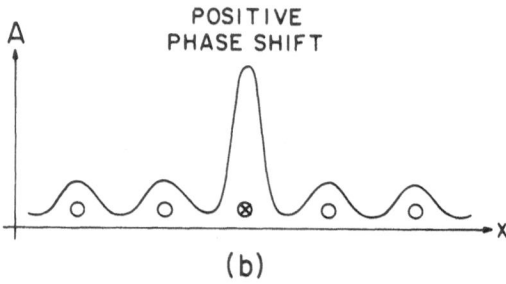

Fig. 2. A schematic of the amplitude A of a wavefunction in the vicinity of an impurity, denoted by the symbol ⊗. The host atoms are denoted by plain circles. Part (a) illustrates a negative phase shift or more repulsive impurity potential, while part (b) illustrates a positive phase shift or more attractive impurity potential.

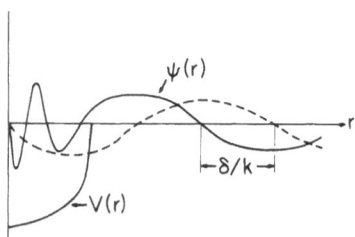

Fig. 3. An illustration of how a potential introduces a phase shift δ where $k = 2\pi/\lambda$ and λ is the wavelength of the unperturbed wavefunction (dotted curve). The unperturbed wavefunction is the state if there were no perturbing potential $V(r)$. The introduction of $V(r)$ changes the wavefunction hear the origin, as illustrated by the solid curve, so that far from the origin it is phase shifted by δ from the unperturbed state.

where the atoms are located at the points $\vec{R}_n$; The Wannier state of ℓ-character near the origin centered about the site $\vec{R}_n$ is denoted by $\phi_\ell(r-R_n)$; $\vec{k}$ is the crystal momentum which distinguishes the state; and ℓ is the angular momentum eigenvalue having the value 0,1,2 for s,p,d states, respectively. Assumption (1) permits neglect of the mixing between states $\psi_{k\ell}$ and $\psi_{k\ell}'$ where $\ell' \neq \ell$. For Ti and $\ell = 2$, the addition of an impurity with no d-character greatly modifies ψ_{k2} in its vicinity. The eigenstates can no longer be characterized by $\vec{k}$ since the impurity scatters a given $\vec{k}$ state into different $\vec{k}'$ states, the same mechanism that produces electrical resistance. In addition, it now costs a lot of energy to have a Wannier state of d-character centered about the impurity which does not like to have this d-character. The result is that in the energy range of interest the Wannier state centered about the impurity has very little amplitude compared to those of the host atoms. This effect is called charging [8]. Fig. 2(a) illustrates this schematically for the case just discussed. Fig. 2(b) schematically illustrates the case of an energy range where only the d-character of the impurity likes to be occupied. This would illustrate a transition metal impurity in a group B host. Numerical estimates indicate that to a good approximation for Ti with group B impurities, the amplitude of the d-Wannier state on the impurity site is very small and can be set equal to zero for our purposes.

The charging effect can be treated in terms of phase shifts [7] from assumption (4). Phase shifts are a convenient way to describe the perturbation introduced by an impurity. The perturbation introduced by an impurity should be localized in a metal because of shielding, so that far enough away from the impurity, the potential and electron states should be of the same type as before the introduction of the impurity. The effect of the impurity in that region is to introduce a phase shift in the wave function away from the impurity as illustrated in Fig. 3. Friedel [7] proved that the phase shift is related to the change in electronic charge around a given site caused by the introduction of an impurity at that site. Z_ℓ, the contribution to this electronic charge change from a given ℓ-character is

$$Z_\ell = \frac{2}{\pi} (2\ell+1) \, \delta_\ell(k_F) \quad , \tag{2}$$

where k_F is the Fermi wave number. By the perfect shielding requirement

$$\sum_\ell Z_\ell = Z \quad , \tag{3}$$

where Z is the difference in number of valence electrons between the impurity and the host.

From our assumption (1), each Z_ℓ acts approximately independently of the others and, for example, the difference in d-valence between the host and solute is accounted for entirely by δ_2. From Eqn. (2), δ_ℓ is determined by

$$\delta_\ell = \pi Z_\ell / (4\ell + 2) \quad , \tag{4}$$

where Z_ℓ is determined from the atomic configuration as given in Table II.

Table II. Valence differences and phase shifts compared to a Ti host

element \ ℓ	Z_ℓ			δ_ℓ		
	0	1	-2	0	1	2
Ti	0	0	0	0	0	0
Ga	1/2	1/2	-2(+8)	$\frac{\pi}{4}$	$\frac{\pi}{12}$	$-\pi/5$ $(\pi-\pi/5)$
Al	1/2	1/2	-2	$\frac{\pi}{4}$	$\frac{\pi}{12}$	$-\pi/5$
Sn	1	1	-2(+8)	$\frac{\pi}{2}$	$\frac{\pi}{6}$	$-\pi/5$ $(\pi-\pi/5)$

As illustrated in Fig. 2, the introduction of the impurity and its corresponding phase shift disturbs the wave function in the vicinity of the impurity. This distortion is appreciable and causes a given Bloch state to scatter into other states and decay with time. Such a scattering causes electrical resistance and is the mechanism for introducing the residual resistance change on alloying. The residual resistance change can be expressed in terms of phase shifts by [9]

$$\Delta\rho = \Delta\rho_0 \sum_\ell (\ell+1) \sin^2 (\delta_\ell - \delta_{\ell+1}) \quad , \tag{5}$$

where

$$\Delta\rho_0 \approx 4\mu\Omega\text{-cm/\% impurity.}$$

Using the values of the phase shifts given in Table II one finds that $\Delta\rho \simeq 10 - 13$ ($\mu\Omega$cm/% impurity) for Ga, Al, and Sn. The measurements [3] give a range from 10 - 19 ($\mu\Omega$-cm/% impurity) in reasonably good agreement with the theoretical estimate. A larger value of $\Delta\rho$ can be obtained from Eqn. (5) if account is taken of the fact that in the solid, Ti has more like a $3d^3 4s^1$ configuration than the atomic one assumed in Table II. It is clear that the simple model can account for the measured values of $\Delta\rho$.

The calculated value of $\Delta\rho$ is dominated by the d-electron contribution. Since the experiments correlate measured values of $\Delta\rho$ with solution hardening we conclude that the d-electrons also dominate in solution hardening, even though the impurity atoms have no d-character in the conduction band. The d-electron contribution occurs just because the impurity atoms have no d-character. The surrounding transition metal host then has to have a major distortion of its d-states to avoid the impurity. The larger the difference Z_2 in the d-valence between impurity and host, the larger is the distortion, and thus δ_2 by Eqn. (4). This increases $\Delta\rho$ as given by Eqn. (5) and by the experimental correlation, also increases the solution hardening.

To complete the theoretical understanding we must explain why increased δ_2 or Z_2 also causes an increased solution hardening. We present an initial start in this direction in the next section.

3. MODEL CALCULATION

In this section we calculate the properties of an isolated impurity in a simple metal. The model consists of a simple cubic lattice with a single impurity at a site which is chosen for convenience as the origin of coordinates. It is assumed that the solid can be treated by the tight binding approximation (TBA) with a single band where the atomic state about each site, including the impurity site, is an s-state. We make a further simplification, the highly localized approximation, assuming the perturbation introduced at the impurity site is localized only on that site [10]. The perturbation can then be described by one parameter, the diagonal matrix element of the perturbation between the atomic (or Wannier state, in general) states centered about the impurity site. We denote this matrix element by E_0. It is convenient to express all energy units in terms of $W/6$, where W is the band width of the pure cubic host, so let $U_0 = 6E_0/W$.

This model has the advantage that analytic solutions can be obtained for all quantities of interest. It has the disadvantage that it does not apply quantitatively to any model of physical interest. The s-symmetry of the model does not make it directly applicable to Ti whose solution hardening properties, we have seen, are dominated by d-electrons. The highly localized approximation does not correctly treat shielding [11]. Yet, this model may give some qualitative trends which are reliable.

First consider the density of states $\rho(E)$ of unit volume of the pure host as shown in Fig. 4. The band is centered about $\epsilon \equiv 6E/W = 0$ where E is the energy of a state. The introduction of the impurity produces a change in this $\rho(E)$ given by $\Delta\rho(E)$, where [12]

$$\frac{W}{6} \Delta\rho(E) \equiv \rho(\epsilon) = \frac{2}{\pi} \frac{d}{d\epsilon} \left(\tan^{-1} \frac{I(\epsilon)U_0}{1-R(\epsilon)U_0}\right) ,$$

$$R(\epsilon) = P\sum_{\vec{k}} (\epsilon-\epsilon(\vec{k}))^{-1} , \qquad (6)$$

$$I(\epsilon) = \pi\sum_{\vec{k}} \delta(\epsilon-\epsilon(\vec{k})) ,$$

P in the definition of $R(\epsilon)$ means principle value, and $\epsilon(k)$ are the eigenvalues of the pure host expressed in units of $W/6$. Plots of $I(\epsilon)$ and $R(\epsilon)$ are given in Fig. 5.

The phase shift introduced by the impurity is given by the expression [12]

$$\delta_0(\epsilon) = \tan^{-1} \frac{I(\epsilon)U_0}{I-R(\epsilon)U_0} . \qquad (7)$$

If we consider a host with one electron per atom (the only choice of an integral number of electrons per atom that leads to a metallic host) the Fermi energy is at $\epsilon = 0$ and the phase shift at the Fermi energy is given by [13]

$$\delta_0(0) = \tan^{-1} I(0)U_0 . \qquad (8)$$

This is plotted as a function of U_0 in Fig. 6. We note that in order for δ_0 to be near unity or larger, as is the case for δ_2 in Table II, $|U_0|$ must be greater than or equal to 1. Thus we are in the region where perturbation theory is not valid and exact solutions such as those presented here are necessary.

We are interested in the energy of the impurity and how it varies under distortion. The energy of the impurity (in units of $W/6$) is, using Eqn. (6), given by

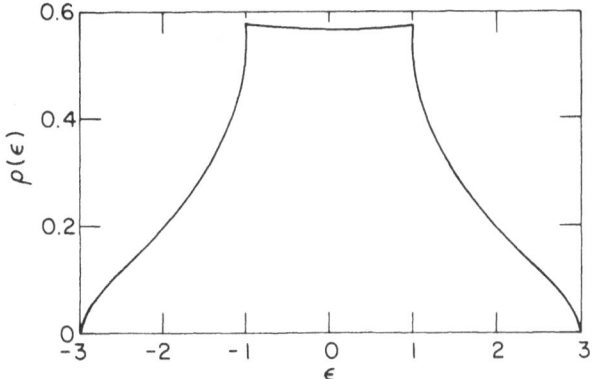

Fig. 4. The density of states $\rho(\varepsilon)$ of a pure solid with a single
s-band and simple cubic lattice measured in energy units of W/6,
where W is the total width of the band.

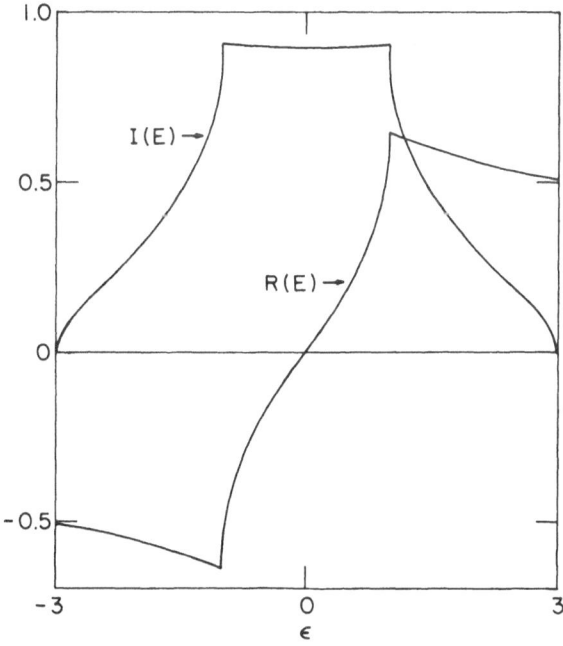

Fig. 5. A plot of $I(\varepsilon)$ and $R(\varepsilon)$ for a pure solid with a single
s-band and simple cubic lattice.

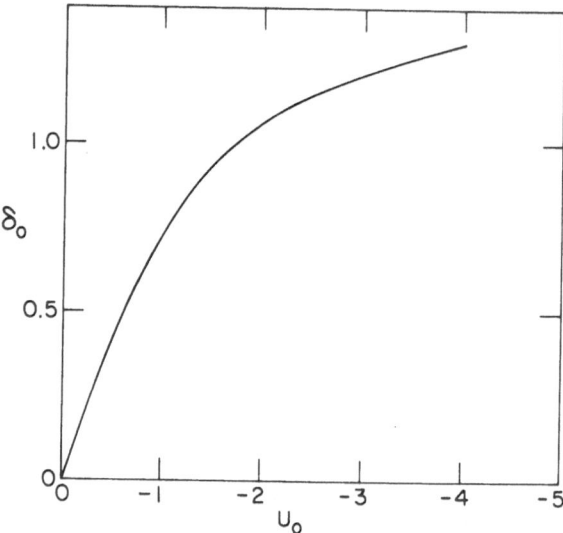

Fig. 6. The phase shift δ_0 introduced by an impurity potential U_0 for the highly localized model.

$$E_i = \int_{-\infty}^{0} \Delta\rho(\varepsilon)\varepsilon \ d\varepsilon = \frac{2}{\pi} \int \varepsilon \frac{d}{d\varepsilon} (\tan^{-1} \frac{I(\varepsilon)U_0}{1-R(\varepsilon)U_0}) \ d\varepsilon$$

$$= -\frac{2}{\pi} \int_{-\infty}^{0} \tan^{-1} (\frac{I(\varepsilon)U_0}{1-R(\varepsilon)U_0}) \ d\varepsilon \qquad (9)$$

The simplest distortion to calculate is a uniform volume change. In such a distortion the only parameter to vary is the lattice constant $\underline{a}$ and its variation is reflected in Eqn. (9) only by a variation of $U_0 = 6E_0/W$. Now U_0 varies with $\underline{a}$ only through the band width W. Differentiating E_i in Eqn. (9) with respect to $\underline{a}$ one finds

$$\frac{dE_i}{da} = \frac{1}{3\pi} \frac{dW}{da} \left[\int_{-\infty}^{0} \varepsilon \frac{d}{d\varepsilon} \tan^{-1} (\frac{I(\varepsilon)U_0}{1-R(\varepsilon)U_0}) \ d\varepsilon \right.$$

$$\left. + \int_{-\infty}^{0} \frac{I(\varepsilon)U_0 d\varepsilon}{(1-R(\varepsilon)U_0)^2+(I(\varepsilon)U_0)^2} \right]$$

$$= \frac{1}{6} \frac{dW}{da} [E_i + \frac{2}{\pi} \int_{-\infty}^{0} \frac{I(\varepsilon)U_0 d\varepsilon}{(1-R(\varepsilon)U_0)^2+(I(\varepsilon)U_0)^2}] \quad . \qquad (10)$$

Since dW/da is dependent only on the host properties and independent of the impurity, the variation of interest is within the bracketed expression on the last line of Eqn. (10).

As discussed above, we are interested in values of $|U_o| \gtrsim 1$. Consider, for example, the case that $U_o \ll -1$. In that case the impurity forms a bound state level below the bottom of the band [12] and we find from Eqn. (10) that

$$\frac{dE_i}{da} \propto \left(\frac{dW}{da}\right) U_o \quad , \tag{11}$$

i.e., the larger the magnitude of U_o the greater is dE_i/da. As discussed in the Introduction, it then follows that the larger the magnitude of U_o the greater the solution strengthening. Using the relation between U_o and the phase shift δ_o displayed in Fig. 6 and the relation between phase shift and residual resistance discussed in Section 2., this result is in qualitative agreement with the experimental results of Collings and Gegel [3].

4.   DISCUSSION AND SUMMARY

The calculations of the previous section gave some justification to correlate the magnitude of the perturbation introduced by an impurity to the solution hardening of this impurity. However the calculation spoke to the question only of a dilation or volume distortion and not to a shear type distortion. Shear type distortions around an impurity are clearly also very important. The calculation of the previous section can straightforwardly be modified to obtain such a result. However, one cannot be assured of even the qualitative correctness of the results for such a model. In shearing one would expect that the difference between s- or d-symmetry of the wave functions would be important, even qualitatively. The directional aspects of bonding between neighboring atoms is quite different in the two cases and this directional aspect should be important in a shear distortion.

To estimate the shear effects one can use an idea due to Collings [14]. If adding impurities narrows the bandwidth of the solid, this brings the properties of the solid closer to molecular binding (which has narrow bands on a band picture). In molecular binding, directional effects are more apparent in the bonding, increasing the shear modulus and thus increasing the hardness of the material. Thus the greater the decrease in the d-band width introduced by a given fraction of impurities, the greater their contribution to solution hardening.

The understanding of d-band narrowing with alloying follows immediately from the discussion of Section 2. The non-transition impurities which introduce large values of δ_2 act, as far as the d-states are concerned, as vacancies. They repel the d-states from their vicinity. This decreases the overlap between the d-states of neighboring atoms and thus decreases the bandwidth [8]. On the other hand, the increased scattering introduced by the impurities tend to broaden the bandwidth. However, it can be shown that the narrowing dominates. A calculation of the band shape changes on alloying non-transition impurities, taking into account all scattering effects [15] is shown in Fig. 7. The narrowing is apparent, and the smearing due to scattering rounds off the sharp structure and lowers the value near the middle of the band.

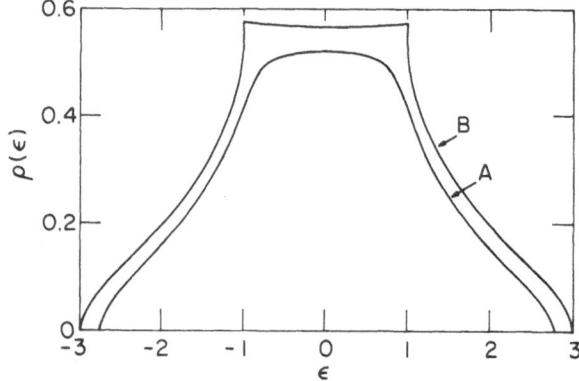

Fig. 7 A plot of the density of states of a random binary alloy consisting of 20% impurities with a strong repulsive potential $U_0 \gg 3$, (curve A). For comparison, the pure solid is given by curve B.

This paper has shown that, at least qualitatively and sometimes quantitatively, the experimental properties of solution hardening can be understood using alloy physics ideas. Arguments are given to justify the view that solution strengthening in Ti alloys is an electronic or local effect where the important interactions occur within an interatomic distance of the impurity. The general idea is that the larger the perturbation of the wave function of the host introduced by the impurity, the greater is the phase shift and the solution hardening. This predicts that one desires impurities which are the greatest number of columns apart from the host in the periodic table in order to obtain the greatest solution hardening. The experimental evidence supports this general conclusion. The paper also outlines a line of calculation, which, if pursued more diligently, should give quantitative agreement with experiments.

A simple experimental check of the ideas of this paper can be made by measuring the change of the various elastic constants of the alloy with concentration. There should be a direct correspondence between the increase of these elastic constants with concentration and the increase of strength of the solid with concentration.

ACKNOWLEDGMENT

The author is pleased to acknowledge several stimulating conversations with Drs. H.L. Gegel and E.W. Collings. This research was supported by a grant from the Air Force Office of Scientific Research.

REFERENCES

1. COTTRELL, A.H., Dislocations and Plastic Flow in Crystals, Oxford, New York, 1953.
2. See, e.g., the discussion by LESLIE, W.C., Met. Trans. $\underline{3}$ 5 (1972).
3. COLLINGS, E.W. and GEGEL, H.L., Paper in this book.
4. See, e.g., HARRISON, W.A., Solid State Theory, McGraw Hill Book Co., New York, pp 203-212, 1970.
5. WATSON, R.E. and EHRENREICH, H., Comments on Solid State Phys. $\underline{3}$ 109 (1970).
6. See, e.g., HARRISON, Solid State Theory, McGraw Hill Book Co., New York, pp 86-87, 1970.
7. FREIDEL, J., Phil. Mag. $\underline{43}$ 153 (1952), and Advan. Phys. $\underline{3}$ 446 (1953).
8. STERN, E.A., Energy Bands in Metals and Alloys, BENNETT, L.A. and WABER, J.T., Eds., Gordon and Breach, p 151, 1968; STERN, E.A., Physics $\underline{1}$ 255 (1965).
9. KITTEL, C., Quantum Theory of Solids, John Wiley and Sons, Inc., New York, p 346, 1964.
10. Some representative references where this model has been used are: SOVEN, P., Phys. Rev. $\underline{156}$ 809 (1967); STERN, E.A., Physics Long Island City, NY, $\underline{1}$ 255 (1965); VELICKY, B., KIRKPATRICK, S. and EHRENREICH, H., Phys. Rev. $\underline{175}$ 747 (1968).
11. STERN, E.A., Phys. Rev. Letters $\underline{26}$ 1630 (1971).
12. CLOGSTON, A.M., Phys. Rev. $\underline{125}$ 439 (1962).
13. RUDNICK, J. and STERN, E.A., Phys. Rev. $\underline{B7}$ 5062 (1973).
14. COLLINGS, E.W., private communication.
15. STERN, E.A. and ZIN, A., Phys. Rev. B, $\underline{9}$ 1170 (1974); ZIN, A. and STERN, E.A., unpublished.

A REVIEW OF SOLUTE EFFECTS ON THE

ELASTIC MODULI OF BCC TRANSITION METALS

E. S. Fisher

Materials Science Division
Argonne National Laboratory
Argonne, Illinois   60439

ABSTRACT

The measured values of the elastic shear moduli C_{44} and C' in bcc transition metals and their bcc solid solutions are reviewed with reference to the influence of the number of electrons in the unfilled d shells. Whereas a classical Fuchs model based on electrostatic and overlap exchange repulsion alone predicts anistropy ratios, C_{44}/C', much greater than unity, measurements show that the band structure contribution causes a smooth but very marked decrease in C_{44}/C' with the e/a ratio increasing from 4.2 to 5.5. From $5.5 < e/a < 6.3$ the C_{44}/C' ratio increases, indicating that the maximum band structure effect occurs at the same e/a range where singularities are observed in other electronic properties of these solid solutions. This behavior makes it possible to predict the influence of solute additions on the individual moduli as well as in their ratio.

In bcc Fe the anistropy ratio is influenced primarily by the degree of ferromagnetic order. H in solution is shown to have a marked influence on the shear moduli of V and this effect is diminished by the precipitation of vanadium hydride. The importance of the band structure contribution relative to that of the exchange repulsion in bcc transition metals is indicated in a comparison of the pressure derivatives of the bulk moduli with those of close packed structures.

199

1. INTRODUCTION

The elastic moduli of most close packed metals and some body centered cubic metals can be understood on the basis of an electrostatic contribution, W_e, a hard core repulsive contribution, W_r, and a Fermi energy term, W_F. The electrostatic term, which is derived from the Coulomb attraction between the positive ions and the negative conduction electron charge is the important long range contribution to the elastic shear moduli whereas the repulsion term, derived from ionic or electron band overlap, is the important short range term in both the shear moduli and in the compressional moduli, where volume changes are involved [1]. The Fermi energy term is evidently an important long range contribution to the bulk modulus of metals and is presumed to be a dominating term in the shear moduli of certain polyvalent metals where the Fermi surface overlaps Brillouin zone boundaries [2-4].

In the case of cubic symmetry there are two principal shear moduli, C_{44} and $C' = 1/2(C_{11}-C_{12})$, where the ratio C_{44}/C' provides a reasonable measure of how well the W_e and W_r terms alone can account for the elastic moduli. In the noble fcc metals, with nearly filled electron shells and relatively small atomic volumes, this ratio ranges from about 1.5 to 3, as is predicted from the Fuchs model for nearest neighbor repulsive forces with a small contribution from the long range forces [1]. For the bcc alkali metals with large atomic volume, this ratio is again extremely large, ranging from 8 to 10 among existing data, and is again consistent with a simple model taking into account primarily the electrostatic long range contribution. As pointed out by Fuchs [1], the addition of W_r terms from interactions between overlapping nearest and next nearest neighbors should tend to make C' vanish and thus make the bcc structure unstable. The fact that bcc structures are in fact found in the metals of Groups V and VI of the periodic table and in the beta-brass type alloys has been discussed by Zener [5] and by Isenberg [6], who propose that W_r forces between second neighbors have to be much more sensitive to interatomic separation than the first neighbor forces. More recently (1969 and 1970) there appeared several experimental studies of the elastic moduli in single crystals of bcc transition metal solid solutions that indicated a very strong dependence of the shear moduli on the electron concentration, or the number of electrons per atom, (e/a) based on the electron configuration in the individual atoms [7,8]. These studies, together with further analysis of earlier data on the unalloyed bcc transition metals, indicated that there is a very large band structure contribution to the shear modulus C' which appears to be independent of interatomic distance and which overwhelms the W_e and W_r contributions. The conclusion from this 1970 data is that the classical Fuchs model, where the short range repulsive forces would be expected to dominate, is probably appropriate at e/a ratios of slightly

greater than 4 and also at e/a in the range of 6.3, but that unique
band structure contributions are very large at intermediate e/a
ratios. A second conclusion from this work [7] is that the magnitude of the C' modulus in a random substitutional bcc solid solution can be predicted on the basis of a rigid band model so that
only the number of d electrons contributed by a solute is relevant.
The purpose of the present review is to consider all the data
available as of September, 1973, in light of the previously published work and to see whether new patterns appear which are more
in line with other properties that vary with a rigid band approximation [9-12] and also more consistent with particular theories that
have been proposed for the band structure contributions to the shear
moduli of metals in general [13,14]. Of special interest here is
the electron transfer model of Fischer et al. [13] which predicts
that in the bcc structures the C_{44} modulus rather than C' modulus
should vary with the filling of the d bands.

The central parameter that will be stressed in this review is
the ratio C_{44}/C'. In addition to the application to the theory of
the elastic moduli, this parameter is also of interest to several
practical applications, where it may be advantageous to have a
material with $C_{44}/C' = 1$. Under such conditions the single crystal
is elastically isotropic and grain boundaries in a polycrystal have
only a minor effect on stress or strain distribution during high
cycle fatigue and, secondly, have only a minimal effect, if any,
on acoustic wave attenuation [15]. It is therefore of practical
interest to be able to adjust alloy composition so as to achieve
the isotropic condition.

In addition to reviewing the shear modulus data, we present
here some information regarding the hydrostatic pressure derivatives
of the bulk moduli. This parameter, dK_T/dP, is of value in estimating the relative importance of the W_r contribution to the elastic
moduli, when compared with other metals [16]. At the present time
this parameter can be deduced from analysis of shock wave data but
we intend here to point out the need for further measurements of
this parameter via ultrasonic measurements on bcc transition metal
alloys.

2. DEFINITIONS AND TECHNIQUES

The single crystal data discussed here are all obtained from
the velocities of ultrasonic waves with frequencies greater than
5 MHz. There are various methods for measuring such wave velocities,
v, all of which are generally accurate to better than 0.5%. Under
these conditions, where

$$v = \sqrt{\frac{C_{ij}}{\rho}} \, , \tag{1}$$

C_{ij} being the appropriate elastic stiffness modulus and ρ being the crystal density, the directly measured moduli are accurate to within 1%. For cubic symmetry all of the elastic parameters of a single crystal, i.e., stiffness moduli, compliance moduli, Young's moduli, Poisson's ratio, bulk moduli, etc., can be computed from a determination of the three principal C_{ij} : C_{11}, C_{12}, and C_{44}. The first, C_{11}, is directly determined from the velocity of a compressional wave along a cube axis whereas C_{44} is the modulus deduced from a shear, or transverse, wave traveling parallel to the cube axis. The determination of C_{12} is slightly more complicated in that it contributes only to waves propagated in other than <100> directions. The most direct procedure to determine C_{12} is from a transverse wave traveling parallel to a <110> direction with atomic displacement in a <1$\bar{1}$0> direction. In this mode of propagation

$$v = v_{T2} = \left\{ \frac{C'}{\rho} \right\}^{1/2}, \tag{2}$$

where $C' = (C_{11}-C_{12})/2$. The subscript T2 arises because the transverse waves for the <110> direction of propagation are not degenerate. Thus v_{T1} is the wave velocity when the ultrasonic transducer is oriented for particle displacement along the cube axis in the {110} plane. In the latter mode

$$v = v_{T1} = (C_{44}/\rho)^{1/2}. \tag{3}$$

The compressional wave velocity for <110> propagation is commonly referred to as

$$v_L = \left\{ \frac{C_{11} + C_{12} + 2C_{44}}{2\rho} \right\}^{1/2}, \tag{4}$$

or $C_L = \rho v_L^2 = C_{11} + (C_{44} - C')$, \tag{5}

where the subscript L stands for longitudinal wave, synonymous with compressional wave.

Eqn. (5) indicates the relation between anisotropy in a cubic crystal and the elastic properties of an isotropic solid. For any direction in a cubic crystal the compressional stiffness modulus deviates from C_{11} by an amount proportional to $C_{44}-C'$. Therefore $C_{44} = C'$ constitutes elastic isotropy for shear as well as for compressional moduli and the parameter $A = C_{44}/C'$ is the commonly accepted measure of anisotropy in a cubic crystal. This quantity also has a rather clear physical implication regarding the central force constants in a bcc lattice, as is seen by looking at a projection of the atomic distribution in {110} type planes (Fig. 1). Given this simple model, the C_{44} type shears are resisted primarily by the nn repulsion, with only a small contribution from nnn atoms. The C' shear resistance has essentially no contribution from nn and is dependent primarily on nnn central forces. It is thus reasonably clear that the high values of A for the disordered beta brass type Cu, Ag, and Au alloys arise primarily from the difference in direct exchange interactions or overlap forces between nn and nnn ions. Similarly large A values in the alkali metals arise from the differences in electrostatic Coulomb forces between the ions and the average negative charge density midway between nn and nnn ions. Thus A values in the order of 10 are the rule for those bcc structures where the cohesion is dominated by either short range exchange repulsion or by the electrostatic terms in a central force model. It follows, then, that for the bcc structure the relative values of C_{44} and C' are a measure of the unusual contributions to cohesion that may arise from the electron band structure or from long range magnetic exchange forces.

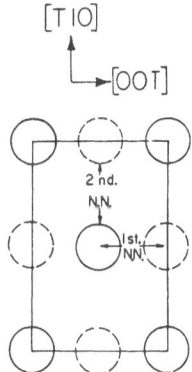

Fig. 1.   The (110) plane of the bcc structure

3.   REVIEW OF DATA FOR SINGLE CRYSTALS OF BCC SUBSTITUTIONAL SOLID SOLUTIONS AND PURE TRANSITION METALS

All of the data available either through publications or private communications are listed in Table I. The listing is based on the e/a ratio for the pure metal or the substitutional solid solution, assuming e/a = 4,5,6, and 7 for the elements of Groups IV, V, VI, and VII, respectively.

Table I. Elastic Moduli of bcc Transition Metals of
Groups IV, V, and VI and Their Solid Solutions, In
Order of Increasing e/a Ratio (at 300 K, with one*
exception)

Composition atom %	e/a electrons/atom	C'	C$_{44}$	C$_{11}$ 10^{12} dynes/cm^2	A	Ref.
*Ti (1000 K)	4.0	0.07	0.336	0.99	4.8	17
Ti 6.98 Cr	4.14	0.124	0.410	1.25	3.3	17
Ti 9.36 Cr	4.19	0.190	0.427	1.331	2.25	17
Zr 9.79 Mo	4.20	0.171	0.367	1.192	2.15	18
Zr 20 Nb	4.20	0.139	0.326	1.164	2.34	19
Zr 25 Nb	4.25	0.161	0.334	1.205	2.08	19
Ti 13.8 Cr	4.28	0.218	0.442	1.399	2.02	17
Ti 28 V	4.28	0.186	0.398	1.411	2.14	20
Zr 30 Nb	4.30	0.177	0.338	1.272	1.91	19
Ti 38 V	4.38	0.241	0.408	1.486	1.69	20
Ti 40 Nb	4.40	0.225	0.396	1.565	1.76	21
Ti 53 V	4.53	0.313	0.413	1.773	1.32	18
Ti 28.4 Cr	4.57	0.325	0.477	1.591	1.47	17
Ti 71 V	4.71	0.448	0.438	2.008	0.98	18
Ti 73 V	4.73	0.407	0.416	1.926	1.02	18
Ti 79 V	4.79	0.431	0.412	1.966	0.96	18
Hf 94 Nb	4.94	0.552	0.293	--	0.53	22
V	5.00	0.546	0.426	2.280	0.78	23
Nb	5.00	0.566	0.284	2.465	0.50	24
Ta	5.00	0.533	0.820	2.650	1.54	25
Ta 4 Nb	5.00	0.537	0.801	--	1.49	26
Ta 8 Nb	5.00	0.547	0.770	--	1.41	26
Ta 1.4 Mo	5.014	0.537	0.824	--	1.53	26
Nb 1.9 W	5.019	0.573	0.292	--	0.510	22
Ta 2.2 W	5.022	0.541	0.831	--	1.54	26
Ta 3.4 Mo	5.034	0.552	0.809	--	1.47	26
Ta 4.3 W	5.043	0.565	0.829	--	1.47	26
Ta 2.3 Re	5.046	0.557	0.839	--	1.51	26
Ta 4.7 Mo	5.047	0.562	0.800	--	1.42	26
Nb 6.7 W	5.067	0.594	0.317	--	0.53	22
Ta 3.8 Re	5.076	0.576	0.843	--	1.46	26
Ta 5.3 Re	5.116	0.598	0.847	--	1.42	26
Nb 16.8 Mo	5.168	0.688	0.295	2.710	0.429	27
V 17.5 Cr	5.175	0.673	0.440	2.549	0.654	18
Nb 23.3 Mo	5.233	0.756	0.320	2.860	0.423	27
Nb 33.9 Mo	5.339	0.895	0.363	3.]57	0.406	27
Nb 51.6 Mo	5.516	1.101	0.626	3.636	0.569	27
Nb 75.2 Mo	5.752	1.347	0.886	4.223	0.658	27
Nb 92.1 Mo	5.921	1.472	1.023	4.540	0.695	27

Table I. (Continued) Elastic Moduli of bcc Transition
Metals of Groups IV, V, and VI and Their Solid Solutions,
In Order of Increasing e/a Ratio (at 300 K)

Composition atom %	e/a electrons/atom	C'	C_{44}	C_{11}	A	Ref.
		10^{12} dynes/cm^2				
Mo	6.00	1.518	1.095	4.661	0.721	8
Cr	6.00	1.413	1.007	3.484	0.713	28
W	6.00	1.594	1.607	5.233	1.008	29
W 2.8 Re	6.028	1.588	--	--	--	22
Mo 7 Re	6.070	1.468	1.148	4.665	0.782	8
W 10 Re	6.10	1.525	--	--	--	22
Mo 16.6 Re	6.166	1.391	1.237	4.650	0.889	8
Mo 26.9 Re	6.269	1.324	1.323	4.607	1.000	8
Mo 35 Re	6.35	1.19	1.35	--	1.13	30

3.1 Relation of C' to e/a

Fig. 2 contains all of the data for the shear moduli C_{44} and
C' of bcc substitutional alloys based on metal solvents of Groups IV,
V, and VI. The original purpose of this plot [7] was to show that
the value of C' is closely related to the number of electrons per
atom and is relatively independent of the period. This conclusion
was a result of studies for a series of metastable bcc Ti-Cr alloys
[17], where the alloy data at 300 K are plotted together with previously measured shear moduli for the unalloyed bcc metals of
Groups V and VI. The average of the measured values at 300 K of C'
for V, Nb, and Ta are similar within 10% and the C' values of Cr,
Mo, and W are within 12% of each other. This was considered rather
good agreement considering the number of different investigators,
differences in techniques and the fact that we chose 300 K as a
reference temperature. The use of 300 K rather than 0 K values is
a result of difficulties in establishing the values of C' at 0 K for
certain of the alloys, where either phase changes or interstitial
gases caused anomalous attenuation problems. Another reason for
the 300 K base is that the most accurate absolute value measurements
are normally made at ambient temperatures, whereas lower temperature
data include errors caused by length changes as well as acoustic
coupling changes. It should also be noted that Cr is in an anti-
ferromagnetic phase at 300 K and that the influence of the magnetic
structure increases with decreasing temperature.

It was found [7] that the Ti-Cr data plus the pure metal data
could be very accurately represented by the equation

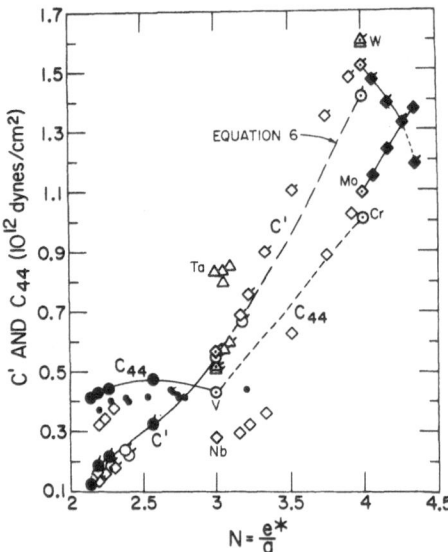

Fig. 2. Measured values from Table I of C_{44} and C' for alloys based on Groups IV, V, and VI transition metals are plotted as a function of e*/a ratio, as defined for Eqn. (6) of text. Primed symbols are for C' and unprimed are for C_{44}. 3d represented by (●,●,○), 4d by (◇◆), and 5d by (△).

$$C' = C'_0 \ (e*/a)^{3.34} \ , \qquad\qquad (6)$$

where e*/a is calculated on the basis of 2, 3, and 4 electrons, respectively, for the Groups IV, V, and VI metals. This equation is represented in Fig. 2 by the indicated line. The e*/a parameter is then an effective e/a ratio and it was conjectured that the physical interpretation may be related to the number of d electrons that contribute to filling of a rigid d band in the range of $2.15 \leq e*/a \leq 4$. Regardless of the physical meaning, the assumption that C' can be reasonably well estimated for any Group IV, V, or VI base metal alloy is fairly consistent with subsequent data, as is indicated in Fig. 2. It is clear that the 3d alloys are most consistent with Eqn. (6), but that even these alloys show some deviations as high as 15%. Such deviations have not been explored with regard to chance causes such as inhomogeneities in composition or effects of interstitial gases. Nevertheless, Eqn. (6) is a reasonably accurate estimate even when the substitutional solute is a transition element higher than Group VI and is not bcc in its own pure metal structure.

As a final comment in regard to the significance of the e/a dependence of C', we note here that C' decreases drastically at e/a <4.15 and at e/a >6.3. The possible significance of these observations to the bcc → ω, α' and α phase changes, at the lower end, and the formation of several complex intermetallic phases at e/a >6.35 are discussed in Ref. [7].

3.2 Relation of C_{44} to e/a and Period

The fact that there is a consistent and systematic relation of C_{44} to e/a was not clear from the initial data for the Ti-Cr alloys. The change in C_{44} with Cr composition is slight, with very little change between the 6.98 at.% alloy, where the pure bcc phase is difficult to maintain by quenching, and the 28 at.% alloy. It should also be noted that C_{44} was measured in unalloyed bcc Ti at temperatures above 1200 K, with values only about 10% smaller than in the 6.98 at.% alloy [17]. The data for the V-Cr alloy indicated that this slight dependence on composition extends beyond V, so that a very marked increase in C_{44} must occur at compositions approaching pure Cr. This relatively complicated mode of C_{44} versus composition behavior and the relatively large differences in C_{44} within pure metals of Groups V and VI suggested that a systematic correlation between a rigid band model and C_{44} was nonexistent. When, however, the data for Zr and Nb base alloys are plotted on the same basis as the Ti-Cr and Ti-V data, it becomes relatively clear that the C_{44} values converge at the Group IV end of the spectrum and both 3d and 4d base alloys are similar with respect to the e/a dependence of C_{44}, i.e., slight dependence at e*/a <3.2 and a very rapid climb between 3.2<e*/a<4.3. At the present time, we can only assume that the 5d alloys have an e/a dependence of C_{44} that is similar to that of the other periods, so that C_{44} in Hf-Ta alloys, for example, is about twice that for the Zr, Ti, V, or Nb base alloys.

3.3 The Elastic Anisotropy Ratio

The above information can be of some value in estimating the modulus defect that may contribute to solution hardening of a particular alloy. It also has a rather important implication in a rigid band model that has heretofore not been stated, namely that the initial band filling primarily affects C', whereas the effect on C_{44} is delayed until e/a corresponds to some concentration between Groups V and VI. Within the higher e/a regime, C_{44} increases at about the same rate as C' but C' decreases at e/a values higher than Group VI whereas C_{44} continues to increase at a rapid rate. Despite these rather complicated differences between C_{44} and C', the net effect of band filling on the A = C_{44}/C' is surprisingly simple, as shown in Fig. 3. The A ratios for the 3d and 4d alloys overlap reasonably well at the left of the diagram, where most of

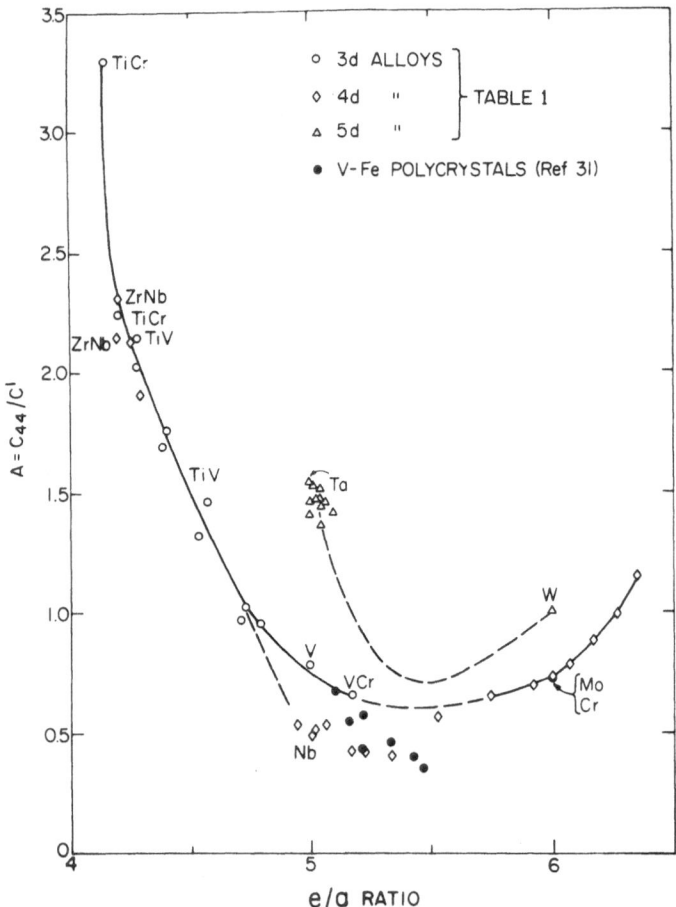

Fig. 3. The anisotropy ratio as a function of e/a ratio for bcc alloys based on Groups IV, V, and VI transition metals. Dashed lines imply a need for further data.

the data exist. These data can be represented with a relatively smooth curve that approaches infinity as e/a→4, but decreases rapidly to a value well below A = 1 before e/a = 5. Because of the relatively small C_{44} for Nb the A ratios of V and Nb are significantly different but it appears that for both 3d and 4d alloys, A passes through a minimum at e/a ∿5.5 and the A versus e/a curves again coincide at e/a approaching 6. The noteworthy features are the relatively smooth U-shaped curve for both 3d and 4d alloys with the minimum in A at about halfway between Groups V and VI. If, as we stated earlier, the high A values at the left and right of Fig. 3

are associated with the classical central force models for the bcc
structure [1], the maximum band structure effect occurs e/a ∿5.5.
This is a somewhat different conclusion than is reached from the C'
data alone, but it does seem to coincide better with certain other
unique features at e/a ∿5.6, which are pointed out in the discussion
(below).

For the 5d alloys the data are indeed insufficient to indicate
that the A versus e/a curve has a maximum at compositions between
Ta and W. It is reasonably certain, however, that C' does in fact
decrease strongly as Hf is added to Ta, whereas C_{44} probably remains
relatively large, so that the A curve probably increases at a more
rapid rate than in the 3d and 4d alloys. On the right side of the
diagram, there are some recent data by Shannette [22] which show
that C' in the W–Re system closely follows the slope indicated by
the Mo–Re data. Thus, we can expect a marked increase in A with a
falling C' and therefore, an A versus e/a curve that is similar in
shape to that for the 3d and 4d alloys.

The relatively consistent results for both 3d and 4d A values
make it clear that the e/a ratio should be a reliable basis for
predicting the elastic anisotropy of an alloy consisting of 3d and
4d transition elements. This application is particularly useful
in the search for alloys that are elastically isotropic, i.e.,
A = 1. The alloys of major interest here are those in the range of
e/a ∿4.7, rather than in the 6.25 range because of the well developed
technology in fabricating the β–Ti alloys. The point of major interest here is that there are in fact two regions of compositions where
isotropic conditions can be achieved.

The recent work of Neighbours et al. [31] is in several respects
of interest to the validity of Fig. 3. This work consisted of compressional and shear modulus measurements, by an ultrasonic technique, in several polycrystalline binary alloys of V–Fe where the
Fe content ranged from 3 to 15 at.%. In order to estimate the A
value of these alloys, the C' modulus was calculated from Eqn. (6)
assuming that each Fe atom contributed 3 electrons to the effective
e*/a ratio and that the grain orientations were completely random.
From the VRH relations [32] for the bulk and shear moduli in polycrystals in terms of C_{11}, C_{12}, and C_{44}, and the calculated C', they
computed C_{44} and the A value for each alloy. These deduced A values
are shown in Fig. 3 as the filled circles. Although the points are
not entirely consistent with the one single crystal point for the
3d alloy to the right of V, they are in line with the data for the
V–Ti single crystals and indicate that the 3d curve may reach a
minimum at approximately the same e/a ratio as the 4d, Nb–Mo, data.
Both the V–Fe and Nb–Mo data indicate that C_{44} and A increases
abruptly at or near e/a = 5.5. A careful study of the variation of
C_{44} over this range of compositions may prove fruitful in defining
the evident close tie between the d band filling and C_{44}.

4. REVIEW OF SINGLE CRYSTAL Fe STUDIES

Although Fe is far removed, in the periodic table, from the Group V and VI transition metals, it also crystallizes in the bcc structure and has the stable bcc structure in its ferromagnetic phase. The influence of dilute solutes on the single crystal moduli for the bcc phase has been reasonably well described [33,35] only for Al and Si, both of which are considered to be magnetically inert solutes. The primary influence of both solutes on the elastic properties appears to arise from the magnetization dilution because of the strong influence of magnetic ordering on the C' modulus. The influence of magnetic order in pure α-Fe on C' and C_{44} is shown in Fig. 4, from Dever's data [36] for the temperature dependence of the two shear moduli. As an initial orientation to the differences between α-Fe and the Groups V and VI transition metals, it should be noted that A = 2.4 at 300 K for α-Fe, indicating greater importance of W_r in Fe. The C' modulus decreases very rapidly with increasing temperature up to the $\alpha \rightarrow \gamma$ transformation temperature, with the greatest slope occurring just prior to the loss of long range magnetic order. In contrast, the C_{44} decrease is almost linear with temperature and is only slightly affected by the magnetic ordering.

The unique relation of C' to magnetization per atom is reflected in the solute dependence of the moduli, shown in Fig. 5. Although Al increases the lattice constant while Si diminishes a_o, their effects on C' are remarkably similar, at least up to 10 at.%. This is as expected from a simple magnetic dilution effect. The C_{44} changes are much less pronounced and the rates of change differ for the two solutes. Because of these small changes in C_{44} any interpretation of the differences on the basis of a_o change would be of little value. For C_{11}, however, the rates of change are quite well defined and the differences between the solute effects seem to correlate with the volume changes. It is clear, however, that volume changes are not the only factors contributing to the C_{11} changes, since dC_{11}/dV is a negative quantity but ΔV for Si additions is also negative. Calculations based on measured values of dC_{11}/dV and ΔV indicate that factors other than size alone are major contributors to the C_{11} and the bulk modulus changes with Al and Si additions.

The primary conclusion from this work is that the ferromagnetic exchange forces in α-Fe have the same kind of effect on the elastic anisotropy as do the d-electrons in the Group V and VI transition metals. This can be seen in Fig. 6, where the A value is plotted against temperature. With the loss of ferromagnetic forces the A value quickly increases to that expected from the Fuch's model con-

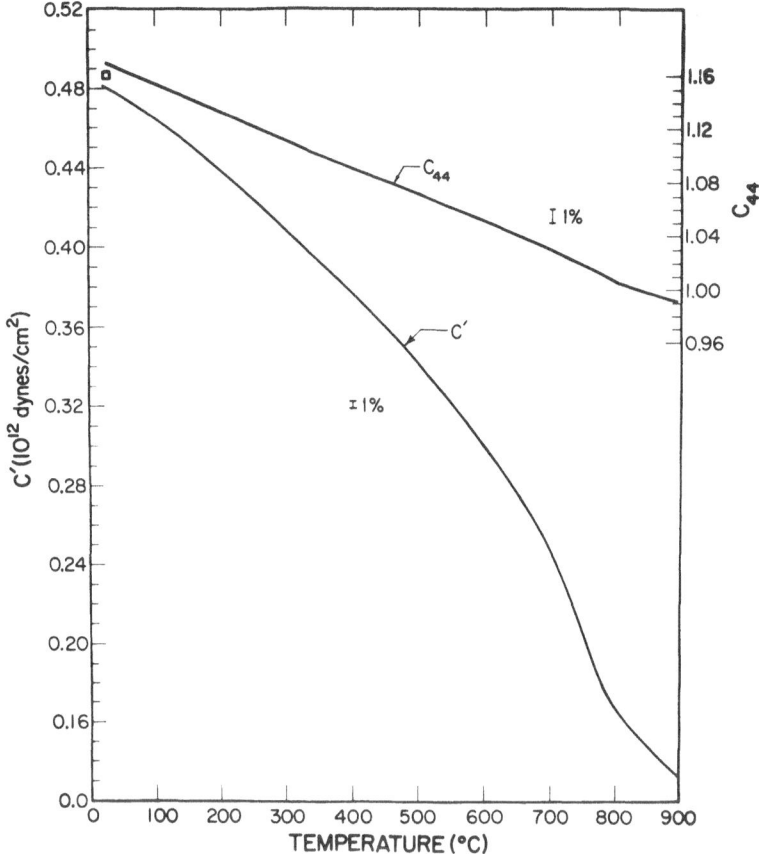

Fig. 4.   The influence of temperature and ferromagnetic order on the shear moduli of α-Fe. $T_c \simeq 767°C$.   (From D. Dever, Ref. [36])

tributions, namely nearest and second neighbor repulsive central forces with electrostatic attraction between the ions and the free electrons.   The influence of alloying with transition metals is known only for polycrystalline Fe samples [37], where C_{44} and C' are almost equally involved.   Single crystal studies of bcc Fe-rich alloys with transition metals would be of great value to test our conclusions from the above studies.

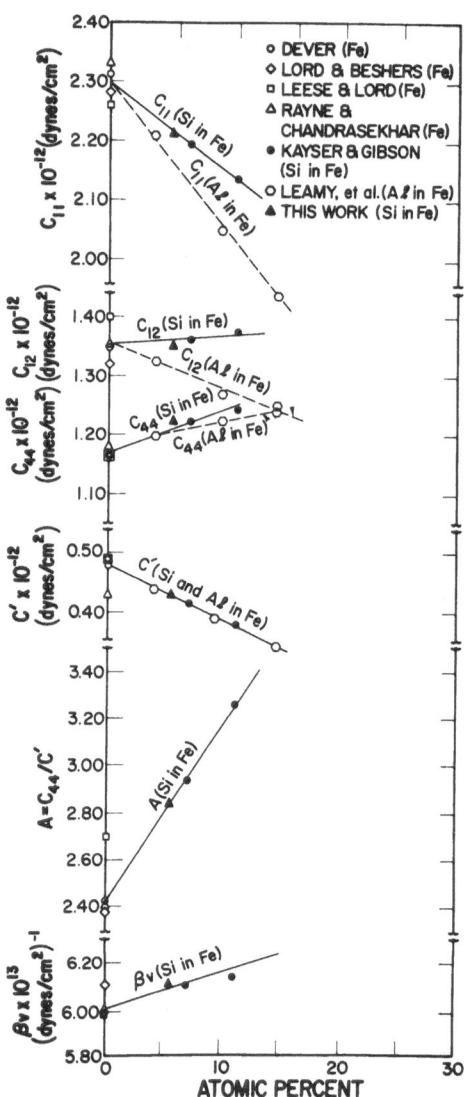

Fig. 5. Effect of Si and Al solute additions on the elastic moduli
of α-Fe. β_V represents volume compressibility. (From J. Routbort
et al., Ref. [35])

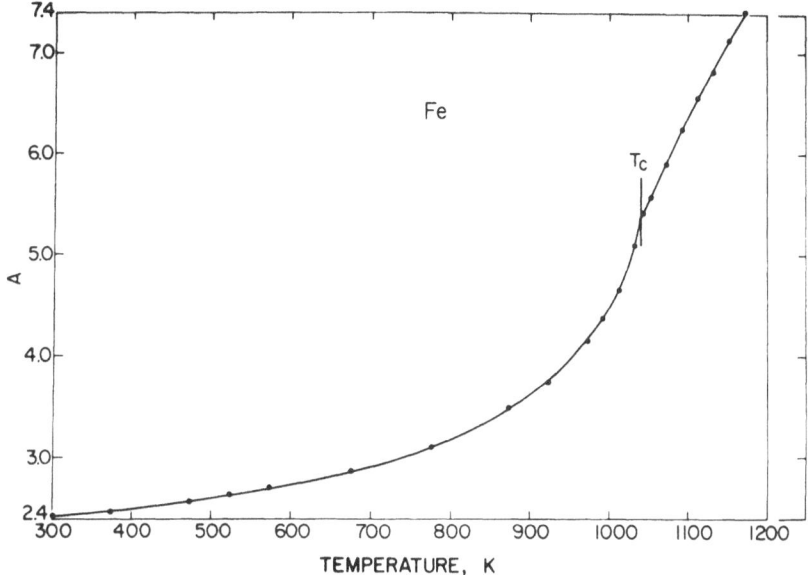

Fig. 6. Influence of temperature and magnetic order on anisotropy
ratio of α-Fe.

5. EFFECTS OF INTERSTITIAL HYDROGEN AND OXYGEN ON V, Nb, AND Ta

There is a large body of literature describing the internal
friction peaks that are related to interstitial solutes in the bcc
transition metals [38]. The anelastic relaxation mechanisms causing
the acoustic absorption can also affect the sound velocities if the
acoustic wave frequencies are less than the thermally activated
jump frequencies. It is not intended here to review the physical
mechanisms associated with movement of the H ions, but to indicate
the extreme sensitivity of the shear moduli to the presence of H
in solution. The studies [39,40] of the effects on the single crystal moduli are in an infant stage, but do show that high frequency
ultrasonics can be a very useful tool for detecting the presence of
the solute, for understanding its effects on the host lattice, and
to indicate the possibility of electronic-charge transfer between
solute H and the host band structure.

The discovery of the "hydrogen anomaly" in the elastic constants
of V was due to Bolef et al. [39] and the timely studies by
Westlake [41] of the solubility of H in V. The anomalies were
clearly associated with the temperatures at which second phase

precipitation begins during cooling. Some more recent [40] work
(Figs. 7,8) shows quite clearly, however, that the anomaly is caused
primarily by the H in solution rather than the presence of a second
phase, and that the solute H causes a large decrease in C' while
increasing C_{44}. The measurements shown in Figs. 7 and 8 were obtained from two different crystals, each of which were deliberately
charged with the indicated H concentrations. The temperature, T_p,
at which precipitation begins increases with H, whereas C' at
$T > T_p$ decreases with increasing H. At $T < T_p$, C' moves toward the
zero-concentration C' value. The measurements shown here suggest
that the hydride precipitate causes an increase in C', but the
increase is relatively small compared to the decrease produced when
H is in solution. The measurements with a second crystal gave similar
results for C' and also show how H influences C_{44}. At $T > T_p$, C_{44}
is clearly increased by the solute, but the effect of H disappears
after complete precipitation. These studies have been extended [40]
to determine whether similar effects occur in the room temperature
values of C' and C_{44} of Nb and Ta, and to determine whether O can
produce similar effects. Table II is a summary of the results described in Ref. [40]. The elastic compliances noted in Table II are
derived from the C_{ij} and are defined as follows:

$$S_{11} = \frac{1}{E_{100}} ,$$

where E_{100} is the Young's modulus in a $< 100 >$ direction.

$$S' = \frac{1}{C'} ,$$

$$S_{44} = \frac{1}{C_{44}} ,$$

and β is the volume compressibility. For all three metals, H causes
significant decreases in C' and E_{100}, only small effects on the bulk
modulus and significant increases in C_{44}. O increases both C' and
C_{44} in V, has essentially no effect on C' for Nb, but causes a large
increase in C_{44} for Nb.

The effect of H on C' appears to be consistent with the Snoek
relaxation effect, where the interstitial occupies either a tetrahedral or octahedral site in the bcc lattice and thus creates a
local tetragonal dipole. The selection rules [42] for the dipole
reorientation due to an induced acoustic wave call for a decrease
in C', or increases in S_{11} and S', and no changes in β and in C_{44}.
These predictions are consistent with Table II if the negative
ΔS_{44} is assumed to arise from some other mechanism and is therefore
independent of $\Delta S'$. We then concluded that H has two different
effects on the elastic moduli, one being the $\Delta C'$ effect associated
with the Snoek relaxation and the second being a ΔC_{44} effect that
may be related to electronic charge transfer.

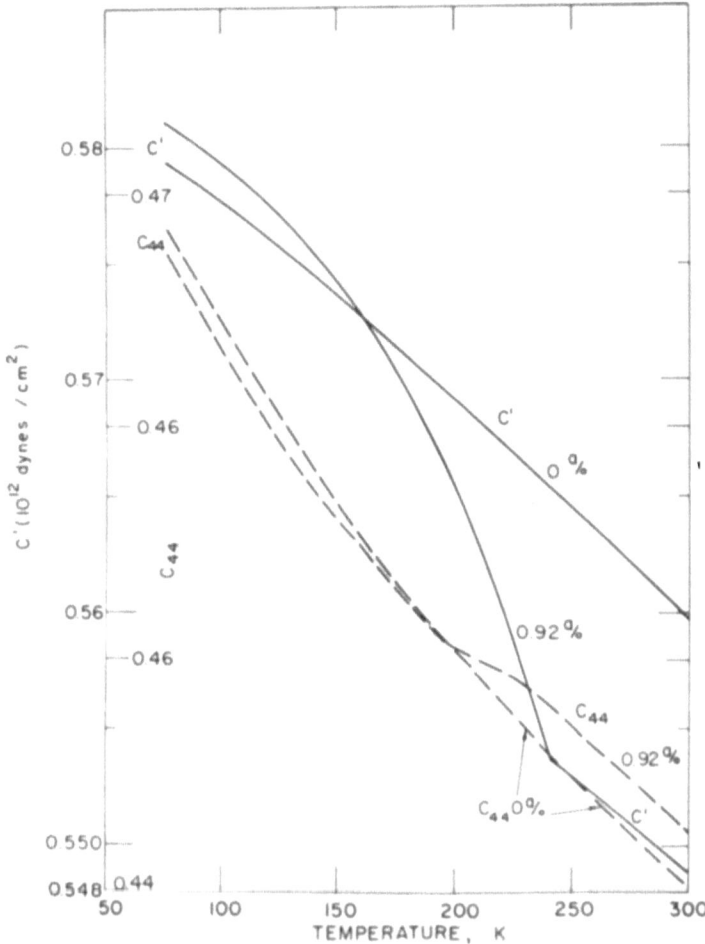

Fig. 7. Influence of hydrogen on C' and C_{44} of vanadium. Abrupt changes in slope occur at hydride precipitation temperatures.

The association of the $\Delta C'$ effect with Snoek relaxation due to noninteracting H solute is consistent with the absence of a negative $\Delta C'$ effect when O is added to dehydrogenated V and Nb. In order to observe a significant Snoek relaxation effect in the modulus it is necessary that the acoustic wave frequency, ω, be of the same order of magnitude, or less, than the thermally activated atomic relaxation frequency, τ^{-1}. With $\omega \sim 10^8$ Hz, when $f \sim 5 \times 10^7$ Hz, and $\tau < 10^{-11}$ sec for H in V, Nb, or Ta at 298 K,

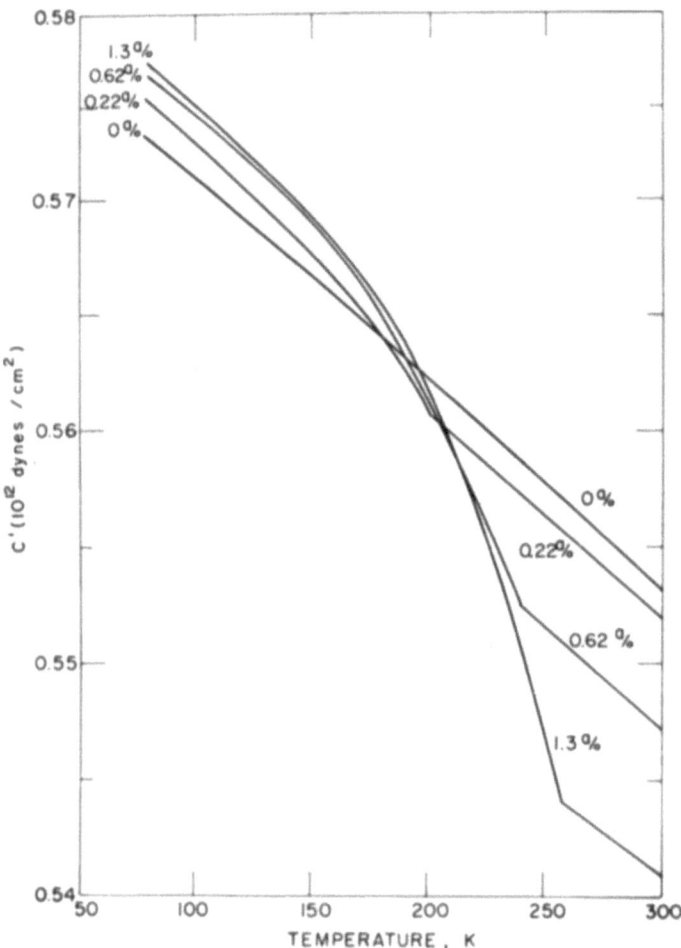

Fig. 8. Influence of hydrogen on C' value in a second single
crystal of vanadium

this necessary condition is fulfilled for H. But for O, with
$\tau \sim 10^6$ sec at 298 K, the $\Delta C'$ effect should not be observed.

At the present time, it is not clear whether the positive ΔC_{44}
effects observed for both H and O are caused by the same mechanism.
The principal uncertainty is because we know that this effect disappears after precipitation of the hydride, but it is not known
whether ΔC_{44} caused by O is solely associated with the solute interstitial.

Table II. Average Relative Modulus Changes per 1 at.%
Solute H or O in V, Nb, and Ta

Crystal	$\dfrac{\Delta S_{11}}{S_{11}}$ %	$\dfrac{\Delta S'}{S'}$ %	$\dfrac{\Delta \beta}{\beta}$ %	$\dfrac{\Delta S_{44}}{S_{44}}$ %
VI-H	1.96	2.16	0.18	-0.54
VII-H	1.60	1.7	0.22	-0.57
VIII-O	-0.54	-0.7	-1.80	-1.30
Nb-H	0.49	0.54	0.08	-1.75
Nb-O	0.00	0.00	0.22	-0.99
Ta-H	0.40	0.43	-0.05	-0.13

6. APPLICATION OF SHOCK WAVE AND HYDROSTATIC
PRESSURE MEASUREMENTS TO THE THEORY OF
COHESION IN THE BCC TRANSITION METALS

The data presented above give clear indications that interatomic
cohesion in the bcc alloys where $2 < e^*/a < 4.3$ does not rely heavily
on the short range interatomic repulsion due to ion overlap.
Attempts have been previously [6,8,29] made to interpret the pure
metal data in terms of two different repulsive energies, W_1 and W_2,
where the subscripts refer to nn and nnn interactions, respectively.
The phonon dispersion curves determined from inelastic neutron
scattering data [43] show quite clearly, however, that such relatively simple models cannot explain the major aspects of the dispersion relations in these alloys, although they are reasonably
successful in the case of the close packed metals and even in α-Fe.
As further proof that the long range forces dominate in these bcc
transition metals, the hydrostatic pressure derivatives of the
elastic moduli may be a useful approach. Although all of the third
order elastic moduli give information regarding the strength of the
interatomic repulsion, the parameter dK/dP, where K is the bulk
modulus, is the least dependent on crystal geometry and is therefore a valid parameter for comparing the influence of short range
forces among various solids.

Assuming a Born-Mayer interatomic repulsive potential

$$W(r) = A \exp \left[-B \left(\frac{r}{r_o} - 1 \right) \right] \ . \tag{7}$$

Where A and B are material parameters, the short range contribution
to K becomes

$$K_{SR} = \frac{2}{3} \cdot \frac{r^2}{\Omega} \cdot \left(\frac{d^2W}{dr^2}\right) \; , \tag{8}$$

where Ω is the atomic volume. The short range contribution to the hydrostatic pressure derivative of K_{SR} becomes

$$\frac{dK_{SR}}{dP} = K'_{SR} = \frac{K_{SR}}{3K_T}(B+1) \quad . \tag{9}$$

If we assume the only other important contribution to K is that from compression of the Fermi electron gas, we obtain

$$K_f = \frac{2}{3}\frac{E_f}{\Omega}Z^* \; , \tag{10}$$

where Z^* is the effective number of free electrons per atom and $K_f = \frac{5}{3}\frac{K_f}{K_T}$, where K_T is the total isothermal bulk modulus. The total pressure derivative for K_T becomes

$$K'_T = K'_f + K'_{SR} = \frac{5}{3}\frac{K_f}{K_T} + \frac{B+1}{3}\frac{K_{SR}}{K_T} = \frac{5}{3} + \frac{K_{SR}}{K_T}\left(\frac{B-4}{3}\right) \quad . \tag{11}$$

Eqn. (11) is useful in that it provides a basis for estimating the factor K_{SR}/K_T which is the relative magnitude of the short range central force contributions to the bulk modulus. If B, the atomic hardness parameter, is assumed to be a constant factor, the value of K'_T should be directly dependent on K_{SR}/K_T. If we make the reasonable assumption for central forces that $K_{SR}/K_T \sim 1/\Omega$, we should find that K'_T for a given metal is related linearly to the initial atomic volume, Ω, through an inverse relation and that at very large Ω, the value of K'_T should approach 5/3.

Before proceeding with a comparison of experimental K' values with Eqn. (11), it is important to mention the methods of deriving experimental values of K'_T. The most direct and most accurate method is to determine the pressure derivatives of the single crystal moduli when the samples are hydrostatically compressed in a liquid or gas medium [16]. For cubic crystals, K'_T is readily evaluated from

$$K'_S = \frac{d}{dP}(C_{11} - 2C_{12})/3 \; , \tag{12}$$

where the S and T subscripts refer to adiabatic and isothermal conditions, respectively. The conversion from K_S' to K_T' is readily accomplished and not a very important factor for most metals at room temperatures.

Since $K_T = -V \frac{dP}{dV}$, the ultrasonic value of K_T determines the initial slope of the volume versus hydrostatic pressure curves that emerge from either pure compression-displacement measurements such as carried out by Bridgman [44] or from X-ray or neutron diffraction measurements with a specimen in a high pressure cell. The K_T' parameter is obviously an important parameter in determining the curvature of the V/V_0 versus P plots and for establishing an equation of state for very high pressure conditions. Although the ultrasonic K_T' value tells us only how the very initial slope varies with pressure, several mathematical derivations of an equation of state have been developed from finite strain theory which allow fairly accurate predictions of V/V_0 versus P plots from a knowledge of ultrasonic K_T' values [45]. The particular one that we are concerned with here is the Birch-Murnaghan equation [45]

$$\frac{P}{K_T} = \frac{3}{2} \left\{ \left(\frac{V_0}{V}\right)^{7/3} - \left(\frac{V_0}{V}\right)^{5/3} \right\} \left\{ 1 - \phi\left(\left(\frac{V_0}{V}\right)^{2/3} - 1\right) \right\} \quad (13)$$

where $\phi = \frac{3}{4} (4 - K_T')$. $\qquad\qquad\qquad\qquad\qquad\qquad\qquad$ (14)

The reason we are concerned with this equation at this point is because there are no direct ultrasonic measurements of K_T' for the bcc transition metals. From V_0/V versus P data, we can, however, derive some K_T' values which we have good reason to believe are fairly close to what will be measured ultrasonically, i.e., they are in good agreement with ultrasonic data where available. To do this accurately with fairly hard metals requires P values that are near to or exceed one megabar. For pressures of this magnitude it is necessary to rely on a third method of measuring equations of state, which is the measurement of high compression shock wave velocities. The equations relating V to P from shock wave and particle wave velocities, respectively U_S and U_p, are readily developed from the equations for conservation of mass, momentum and energy during high velocity impact [46]. The techniques for measuring U_S and U_p have been developed to a very sophisticated level and data for the bcc transition metals have been obtained at pressures up to 1.75 M bar [46]. The equations for deriving the V_0/V from the U_S and U_p values at a given pressure, P, are given in Ref. [46]. The appropriate value of $(K_T')_0$ is then easily

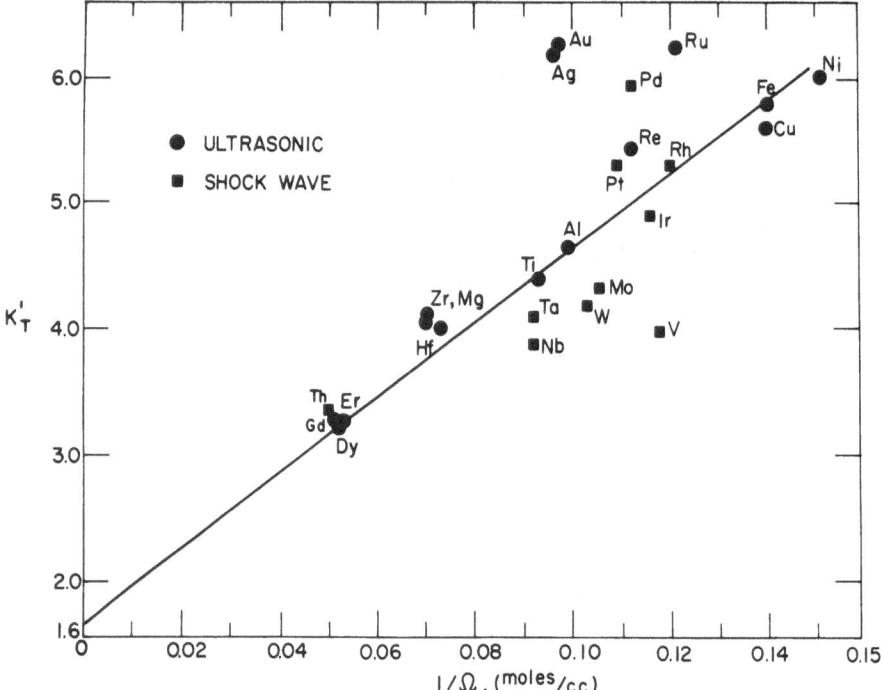

Fig. 9. Plot of the pressure derivative of the isothermal bulk modulus in several metals, as a function of molar volume at zero pressure and 300 K. Note that K_T' values for bcc transition metals are relatively small and not consistent with Eqn. (11).

evaluated for a given P from Eqn. (13). For those cases where pressure-induced phase changes do not occur, the maximum deviation in $(K_T')_o$ values with P is within 20% over $0 \leq P \leq 1.5$ Mbar.

A comparison between the K' values derived by the equation-of-state method with those determined from ultrasonic data obtained under hydrostatic pressure shows that the agreement is within 10% where reliable data are available. It appears reasonable to assume that, in the absence of direct ultrasonic data for the K' of bcc metals, the K' values derived from the shock wave data are valid to within 10%.

Fig. 9 is a comparison of the experimental values of K_T' (as evaluated from ultrasonic and shock wave data) with Eqn. (11). The straight line drawn in the figure connects the value K' = 5/3 with the ultrasonic K' values for the hexagonal close packed rare earth metals, Gd, Dy, and Er. This line suggests that linear K' <u>versus</u> $1/\Omega$ plot is indeed consistent with the small core fcc and hcp metals and nearly consistent with many of the noble metals. The obvious exceptions are those close packed metals where K' is much too large, Ag, Au, Pd, and Ru and the bcc transition metals where the K' values are too small. It is important to note that K' $\geq$ 5 seems to represent those fcc and hcp metals where an important short range contribution is to be expected and α-Fe is also included in this group. The bcc transition metals of Groups V and VI are clearly different in their K' values and in the relative importance of the short range contribution to the elastic moduli.

7. DISCUSSION AND CONCLUSIONS

Although there remain many areas where experimental data are lacking, it appears that the elastic moduli of most bcc transition metal solid solutions can be predicted to within 15% by assuming that the (s + d) e/a ratio varies with the position of the solvent and solute transition elements in the periodic table. In this context the bcc structure can only be retained at ambient temperatures with $4.2 < e/a \lesssim 6.4$ and in certain Fe-base solid solutions where long range ferromagnetism is found. In the latter case, the value of the C' elastic modulus is evidently directly dependent on the influence of composition on the magnetization per atom or long range ferromagnetic order.

The explanation for the composition dependence of the elastic moduli in the range of 4.2<e/a<6.4 is evidently closely related to the band structure properties which also cause the cyclic rigid band behavior of the electronic heat capacity coefficient [47], γ_e, the nuclear-lattice relaxation time parameter [10] $(T_1T)^{-1/2}$, the N.M.R. Knight shift [11], the paramagnetic susceptibility [48], and the superconducting transition temperature [9], T_c. All of these data show singularities at e/a∿5.6, where γ_e, $(T_1T)^{-1/2}$, T_c and χ, have a minima and K_V (the Knight shift for the V atom in several V alloys) has a maximum value. All of these singularities suggest that the densities of electron states at the Fermi level attain a minimum value at this range of e/a for all three periods, 3d, 4d, and 5d. In addition to the above parameters that are normally associated with the band structure, Jones [12] has noted that the extent of solid solubility of H in Ti-Mo alloys also decreases to essentially zero at e/a of 5.6, remains zero up to e/a = 6.0 but then increases with addition of Re to Mo. It seems, then, that the band structure singularity in the region of 5.5<e/a<6.0 is the cause of the abrupt change in the composition dependence of C_{44} and the minimum in the anisotropy ratio A = C_{44}/C'. The composition dependence of C', by itself, is also dependent on the band structure but not specifically with the e/a∿5.5 singularity.

The composition dependence of C_{44} is qualitatively consistent with an electron-transfer model proposed by Leigh [3] for explaining the elastic anisotropy in Al. In this model, the total Fermi energy changes with a shear distortion when the number of electrons in the conduction bands exceeds the capacity of the Brillouin zone. A negative contribution to C_{44} will arise because the overlap across the hexagonal faces of a cubic crystal is increased by a C_{44} distortion. The magnitude of the negative contribution to C_{44} will in turn depend on the density of electron states in the partially filled overlapping band. Thus the total band contribution to C_{44} should contain two terms, which are written as follows according to the development of Fischer et al. [13]

$$C_{44} = \sum_{ij} N_{ij} \left(\frac{\partial^2 E}{\partial \gamma^2}\right)_0 - \frac{1}{\mu_0^2}\left(\frac{\partial E}{\partial \gamma}\right)_0^2 \chi(T) \quad , \tag{15}$$

where the first term is the second derivative of the average band energy, E, with respect to strain γ, and is proportional to the number of electrons in the full band N_{ij}. The second term arises from the change in energy due to the electron transfer and is proportional to the spin susceptibility, χ, divided by the magnetic moment per electron. With a relatively high band density of states, as is the case for e/a<5.5, C_{44} should be diminished by the overlap of the Fermi surface across the reciprocal (111) faces. Thus the electron transfer may account for the relatively small change in C_{44} with e/a<5.5. The relatively rapid increase in C_{44} at e/a>5.5 evidently implies that the overlap bands are filled and thus electron transfer cannot occur. The positive contributions to C_{44} with increasing e/a are present either because of the shift in energy surfaces with zone distortion, as proposed by Leigh [3], or because of the increase in the d electron exchange repulsion with the decrease in lattice constants.

The rapid increase of C' with e/a between 4.2<e/a<6 cannot be explained by any proposed electron transfer model, although it seems to be closely related to the addition of electrons to unfilled d bands. In view of the importance of second nearest interactions to C', it appears that the directional nature of the d band overlap repulsion is responsible for the smooth increase of C' with e/a. The decrease of C' at e/a>6 is consistent with the increase in nearest neighbor overlap forces, and C_{44}, at e/a>5.5. Although this exchange repulsion is not as strong as in the fcc metals, as is indicated in the dK/dP plot of Fig. 9, it evidently increases very rapidly at e/a>6 so as to cause the rapid increase in the A ratio.

Other evidence for the strong dependence of the elastic moduli of these bcc alloys on the band structure effects is indicated in the unique dependence of the temperature coefficient of C_{44} on the e/a ratio of Nb alloys [27], where the algebraic sign changes from an abnormal positive value at e/a∿5.2 to a normal negative value at e/a∿5.5. As pointed out in Ref. [13], this behavior is consistent with the electron transfer model given by Eqn. (15).

ACKNOWLEDGMENT

The author is indebted to M. H. Manghnani and K. Katahara of the Hawaii Institute of Geophysics for their cooperation in the preparation of Fig. 9 of this paper. This work was carried out under the auspices of the U.S. Atomic Energy Commission.

REFERENCES

1. FUCHS, K., Proc. Roy. Soc. (London) <u>A151</u> 585 (1935) and Proc. Roy. Soc. (London) <u>A157</u> 444 (1936).
2. JONES, H., Phil. Mag. (7) <u>43</u> 105 (1952).
3. LEIGH, R.S., Phil. Mag. <u>42</u> 139 (1951).
4. REITZ, J.R. and SMITH, C.S., Phys. Rev. <u>104</u> 1253 (1956).
5. ZENER, C., <u>Elasticity and Anelasticity of Metals</u>, Univ. of Chicago Press, Chicago, 1948.
6. ISENBERG, I., Phys. Rev. <u>83</u> 637 (1951).

7. FISHER, E.S. and DEVER, D., Acta Met. $\underline{18}$ 265 (1970).

8. DAVIDSON, D.L. and BROTZEN, F.R., J. Appl. Phys. $\underline{39}$ 5768 (1968).

9. HULM, J.K. and BLAUGHER, R.D., AIP Conf. Proc. No. 4, 1 (1971).

10. VAN OSTENBERG, D.O., SPOKAS, J.J. and LAM, D.J., Phys. Rev. $\underline{139}$ A713 (1965).

11. BERNASSON, M., DESCOUTS, P., DONZE, P. and TRAYVARD, A., J. Phys. Chem. Solids, $\underline{30}$ 2453 (1969).

12. JONES, D.W., PESSALL, N. and McQUILLAN, A.D., Phil. Mag. $\underline{6}$ 455 (1961).

13. FISCHER, O., PETER, M. and STEINEMANN, S., Helv. Phys. Acta $\underline{42}$ 459 (1969).

14. PASTERNAK, M., STEINEMANN, S. and PETER, M., Helv. Phys. Acta $\underline{41}$ 1296 (1968).

15. SMITH, R.T. and STEPHENS, R.W.B., Progress in Applied Materials Research, STANFORD, E.S., FEARON, J.H. and McGONNAGLE, W.J., eds., Temple Press, Vol. 5, p 41, London, 1964.

16. FISHER, E.S., MANGHNANI, M.H. and KIKUTA, R., J. Phys. Chem. Solids $\underline{34}$ 687 (1973).

17. FISHER, E.S. and DEVER, D., Science, Technology, and Application of Titanium, JAFFEE, R. and PROMISEL, N.E., eds., Pergamon Press, p 373, 1970.

18. FISHER, E.S., to be submitted for publication.

19. GOASDOUE, C., HO, P.S. and SASS, S.L., Acta Met. $\underline{20}$ 725 (1972).

20. GRAHAM, L. and ALERS, G., private communication.

21. REID, C.N., ROUTBORT, J.L. and MAYNARD, R.A., J. Appl. Phys. $\underline{44}$ 1398 (1973).

22. SHANNETTE, G.W., private communication.

23. ALERS, G.A., Phys. Rev. $\underline{119}$ 1532 (1960).

24. GRAHAM, L.J., NADLER, H. and CHANG, R., J. Appl. Phys. $\underline{39}$ 3025 (1968).

25. LEISURE, R.G., HSU, D.K. and SEIBER, B.A., J. Appl. Phys. $\underline{44}$ 3394 (1973).

26. ARMSTRONG, D.A. and MORDIKE, B.L., J. Less-Common Metals $\underline{22}$ 265 (1970).

27. HUBBELL, W.C. and BROTZEN, F.R., J. Appl. Phys. $\underline{43}$ 3306 (1972).

28. PALMER, S.B. and LEE, E.W., Phil. Mag. 24 311 (1971).

29. FEATHERSTONE, F.H. and NEIGHBOURS, J.R., Phys. Rev. $\underline{130}$ 1324 (1963).

30. SMITH, C.S., unpublished data.

31. DONEGAN, J.J. and NEIGHBOURS, J.R., Acta Met. $\underline{21}$ 821 (1973).

32. HILL, R., Proc. Phys. Soc. (London) $\underline{A65}$ 349 (1952).

33. LEAMY, H.J., GIBSON, E.D. and KAYSER, F.X., Acta Met. $\underline{15}$ 1827 (1967).

34. KAYSER, F.X. and GIBSON, E.D., Ames Laboratory Report IS-1500 (1966).

35. ROUTBORT, J.L., REID, C.N., FISHER, E.S. and DEVER, D.J., Acta Met. $\underline{19}$ 1307 (1971).

36. DEVER, D., J. Appl. Phys. $\underline{43}$ 3293 (1972).

37. SPEICH, G.R., LESLIE, W.C. and SCHWAEBLE, A.J., Met. Trans. $\underline{3}$ 2031 (1972).

38. WERT, C.A., J. Phys. Chem. Solids 31 1771 (1970).
39. BOLEF, D.I., SMITH, R.E. and MILLER, J.G., Phys. Rev. B3 4100 (1971).
40. FISHER, E.S., WESTLAKE, D.G. and OCKERS, S.T., phys. stat. sol. 28 (1975).
41. WESTLAKE, D.G., Phil. Mag. 16 905 (1967).
42. NOWICK, A.S. and HELLER, W.R., Adv. in Physics 12 251 (1963).
43. WOODS, A.D.B., AIP Conf. Proc. 4 45 (1972).
44. BRIDGMAN, P.W., Proc. Am. Acad. Arts Science 32 83 (1953).
45. BIRCH, F., J. Geophys. Res. 57 227 (1952).
46. McQUEEN, R.G., MARSH, S.P., TAYLOR, J.W., FRITZ, J.N. and CARTER, W.J., in High Velocity Impact Phenomena, KINSLOW, R., ed., Academic Press, New York, p 293, 1970.
47. HEINIGER, F., BUCHER, E. and MULLER, J., Phys. kondens, Mater. 5 243 (1966).
48. CHILDS, B.G., GARDNER, W.E. and PENFOLD, J., Phil. Mag. 5 1267 (1960).

USE OF THE FRÖHLICH-RAIMES AND OTHER ELECTRONIC MODELS

FOR THE PHYSICAL PROPERTIES OF METALS AND ALLOYS

James T. Waber

Department of Materials Science
Northwestern University
Evanston, Illinois 60201

ABSTRACT

The effectiveness of the Fröhlich-Raimes model in treating force constants and alloy behavior is discussed. Typical predictions of trends which have been made relate to: the deviation of Vegard's Law and the variation of the bulk modulus with pressure and with alloying elements. The simple model of Koskimaki and Waber using the linear combination of the density of states has led to several useful predictions for Ti-based alloys. This model will be discussed and recent results will be reviewed.

1. INTRODUCTION

Several years ago Brooks [1] and Wigner and Seitz [2] took up an investigation of the effectiveness of a volume-dependent theory of metals originally developed by Fröhlich [3] in 1937 and modified by Raimes [4,5] to include several conduction electrons per atom. In an unpublished thesis, Bernstein [6] applied this Fröhlich-Raimes (FR) method to rare earth metals. Brooks notes that the equilibrium lattice spacing in normal metals is essentially determined by pressure of the Fermi gas of free electrons.

In 1961, Larson and Waber [7] reinvestigated the problem and noted that an asymptotic singularity existed in the basic equation which had not been discussed. When this was removed by retaining more terms before truncation of a series expansion for the wave function, better agreement with experimental data was obtained.

Another modification by Waber, Larson, and Smith [8] was to derive the equations assuming that the state at the bottom of the conduction band could have an angular momentum greater than $\ell = 0$. This permitted them to investigate transition metals. A brief review of the theory and some results will be given below before applying it to alloys, since the recent developments have not been reported in the literature.

Before continuing with that topic, it is desirable to give a little background to the second model which appears to be unknown except to a small circle of colleagues. In the case of metals, the Pauli exclusion principle requires that a distinct set of quantum numbers be assigned for each conduction electron coming from an atom in the solid. Instead of the usual set of n, ℓ, m, and spin for free atoms, they are k_x, k_y, k_z, and spin, and are associated with the momentum $\underset{\sim}{k}$ of itinerant electrons. If the density of atoms per cc becomes large enough, the discrete spectrum of energies associated with each k value can be replaced by a piecewise continuous E(k) curve. The number of discrete states within any energy range E to E + dE is replaced by the density-of-states function N(E). The band structure of metals is concerned with calculating such N(E) curves for a pure metal and with deducing various observable quantities from them.

When the original proponents of this theory, such as Jones [9] and Hume-Rothery [10], applied it to metallic alloys, they assumed a Rigid Band model (RB). Slater [11] made one of the first uses of this model to explain the variation of magnetism of alloys; the famous Slater-Pauling curve [12] was the result. These men were keenly aware of the limitations of the model, but insufficient information was available for them to regard it initially as anything but a working hypothesis. It has proved to be very successful in making important predictions during the last 35 years, and many experimentalists have verified the essential features. The comments of Raimes [13] are interesting in this regard. However, small and sometimes large deviations from the RB prediction have been repeatedly noted.

In an extensive series of band calculations for 3d transition metals, Waber and Snow [14] showed that the band structure was far from rigid for such a series of metals; but they also were able to show why the RB model worked well. Despite changes in the shape and the location of the bands for individual metals, the plot of $N(E_F)$ at the Fermi level E_F when plotted against the atomic number had the two-hump shape of the original E(k) curve computed for pure metals by Manning [15] and by Chodorow [16].

Waber proposed that it would be preferable to use a linear combination of the two (or three) densities of states for the constituents in dealing with alloys. In discussions during the conference on the

Electronic Density of States held at the Bureau of Standards, Waber used the LCDS model to estimate [17] (a) density-of-states curves which have three humps, and (b) a typical curve of $N(E_F)$ <u>versus</u> composition for one type of Ti alloy. He also presented the three peak $N(E)$ curves for several 4d metals. Koskimaki and Waber [18] have computed these for a number of alloy systems. Collings and Gegel [19] have made effective use of the calculations recently. It is appropriate to present further the justification of this LCDS model and to record some of the results obtained with it. Independently, Stern [20] studied a form of the LCDS model and concluded that E_F should be constant.

Let us contrast these two approaches to the behavior of alloys. They depend very directly on the average spatial density of electrons in the metal. The modifications of the Fröhlich-Raimes Theory ignore all questions about the arrangement of atoms in the solid. It is also assumed that the Fermi energy ($\hbar^2 k^2_F$) depends only on the number of electrons per atom (i.e., on the (e/a) ratio).

The LCDS theory is predicated on the detailed $E(k)$ curves for each constituent, which in turn is based on the local arrangement of atoms. That is, Snow and Waber [14] have shown that the $E(k)$ curve is not the same for the bcc and fcc forms of the same metal.

Thus, the LCDS model is better able to cope with alloys and such phenomena as short range order and clustering than is the volume-dependent FR model. It in turn is more effective than the (e/a)-dependent RB model. Each model has a range of problems for which it is effective. For the calculation of cohesive energy, compressibility, etc., the FR approximation will be very useful. For the behavior of the specific heat and magnetism of alloys, the LCDS model will be a distinct improvement over the RB model. In principle, the volume (or pressure) dependence of $E(k)$ curves can be calculated by a band calculation such as the APW method. Averill [21] and, independently, Perrot [22] have calculated the cohesive energy for alkali metals with considerable success. Very recently, Snow [23] has done it for Cu. However, these were extensive calculations which consumed large amounts of time on the computer. It is not practical to do exploratory research on alloys by such methods.

The two approximations outlined, in contrast, are very simple to carry out and therefore lend themselves to treating complicated problems. It is with these justifications and limitations in mind that we undertake to study the behavior of alloys.

2. FORMULATION OF FRÖHLICH-RAIMES MODEL

The basic assumption is that the behavior of the conduction electrons in a metal closely approximates those confined to a sphere of radius R_s. Next, one assumes that the ion core with its charge of Ne (or N, in the atomic units where $e = \hbar = m = 1$) does not occupy a significant fraction of the sphere. One solves the Schroedinger equation for one electron confined to the Wigner-Seitz sphere, which also contains an ion core. We will call $f_{\ell o}(r)$ the radial part of the wave function in this field (averaged over the 4π solid angle). The subscript zero indicated the ground state for the angular moment ℓ. The Schroedinger equation can be written as

$$\frac{df_{\ell o}}{dr^2} + \frac{2}{r} \frac{df_{\ell o}}{dr} + \left[E_o(R_s) - V_\ell(r) - \frac{\ell(\ell+1)}{r^2} \right] f_{\ell o} = 0 \quad . \quad (1)$$

The usual Wigner-Seitz boundary condition [24] is applied, namely, that the gradient vanishes according to

$$\frac{df_{\ell o}}{dr} \bigg|_{r=R_s} = 0 \quad . \quad (2)$$

In the space available, it is not practical to discuss the mathematical part of this problem. Only the results in the form of the principal equations will be indicated.

As Waber, Larson, Adachi, and Smith [25] show, the lowest eigen-value E_o depends on R_s in the following simple way

$$E_o(R_s) = \frac{N}{R_s} \left(\frac{R_o^2}{R_s^2} - 3 \right) + \frac{3\ell(\ell+1)}{R_s^2} \left(1 - \frac{2R_o}{3R_s} \right) \quad . \quad (3)$$

The parameter R_o corresponds to the radius at which the energy becomes a minimum. At the other extreme end, when R_s becomes infinitely large, E_o approaches I_n, which is the nth ionization potential for removing all but one of the valence electrons from the atom in question.

Let us call β^2 the positive energy difference between the free ion and the E_o for the metallic case, and call the ionization potential w^2. After substituting R_o for R_s in Eqn. (3), we find that

$$\beta^2 = - E_o(R_o) - I_n \quad (4)$$

$$= \frac{2N}{R_o} - \frac{\ell(\ell+1)}{R_o^2} - w^2 \quad . \quad (4a)$$

In the procedure outlined initially by Fröhlich, one expresses the eigen-function in terms of a power series

$$f_{\ell o}(r) = \exp\,(-\,wr)\,\,r^P \sum_{k=0}^{J} C_k\,r^{-k} \quad , \tag{5}$$

where $P = n/w$.

The recursion formula for the coefficients in the power series can be shown to be

$$\frac{C_{k+1}}{C_k} = \frac{(P-k-1)(P-k-2)\,-\,\ell(\ell+1)}{-\,2w(k+1)} \quad . \tag{6}$$

Then one requires that the function be well behaved at the special value R_o of the sphere boundary. Thus, as a generalization of the solution which Fröhlich and Raimes found, one can write

$$\frac{1}{f_{\ell o}}\,\frac{df_{\ell o}}{dr}\bigg|_{r=R_o} = \frac{R_o(\beta-A)\,\cot\,\beta R_o\,-\,(1+\beta R_o A)}{R_o\,(1+Ar_o\cot\,\beta R_o)} \quad , \tag{7}$$

where A is an arbitrary small constant (which it will be convenient to set to zero) [25].

Substitution of the wave function alternatively leads to

$$\frac{d\,\ell n\,f_{\ell o}(r)}{dr}\bigg|_{r=R_o} = \frac{\sum\limits_{0}^{J}\,(-k)C_k/r^{k+1}}{\sum\,C_k r^{-k}}\bigg|_{r=R_o} \quad , \tag{8}$$

which can be used to evaluate the right-hand side of Eqn. (7). By experience, one need include at most six terms, and in many cases, two will suffice. When $\ell = 0$, these equations reduce to the ones given by Raimes.

A plot of the ground state radius R_0 is presented in Fig. 1, as a function of I_4. A discontinuity is found because the cotangent function approaches $(\pm\infty)$ when its argument equals π. This behavior of the logarithmic derivative corresponds to $f_{\ell o}$ vanishing at the cell boundary for a specific energy. If one desires, this difficulty can be eliminated by retaining the terms multiplied by A in the logarithmic derivative, as the smooth curves show. Insets for other values of N are shown.

The next step in this volume-dependent theory is to put N-1 free electrons back into the spherical cell. There are several types

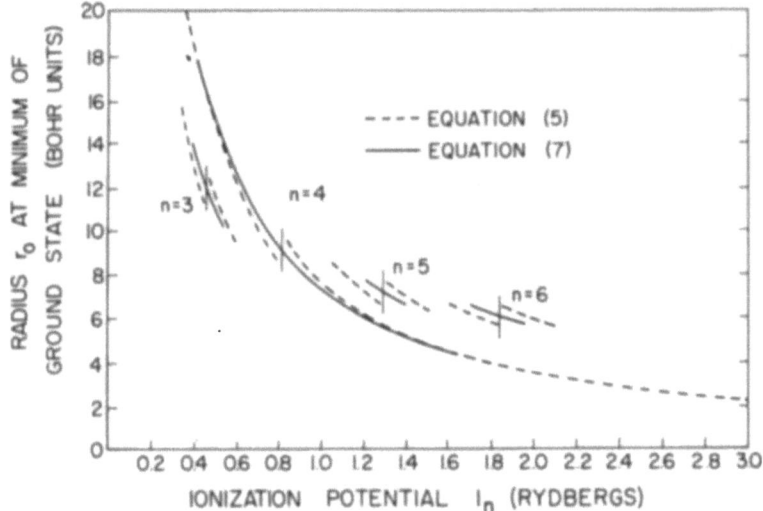

Fig. 1. Plot illustrating the dependence of R_O on the ionization potential w^2 and on the valence N. The continuous curve is band N=4. Waber and Larson used r_o in place of R_O. Their Eqn. (7) corresponds to Eqn. (6) herein, and their Eqn. (5) is obtained by setting A=0.

of interactions between the electrons. The first is Coulomb energy of the N electrons. The second is the mean Fermi energy of these electrons confined to sphere of radius R_s. The third is the free electron exchange energy. A small correction is also included for correlation. Thus, the total energy for the N valence electrons is given by the simple formula

$$E(R_s) = E_o(R_s) + \frac{1.2N}{R_s} + \frac{2.21}{r_s^2} - \frac{0.916}{r_s} - E_{corr} \quad . \tag{9}$$

Inherent in this expression is the idea that $f_{\ell_0}(r)$ is approximately constant throughout a significant fraction of the sphere (for the cases studied by Raimes [4], the fraction was approximately 90%). In a uniform electron gas, we can set r_s as the radius of the Fermi hole, namely, $R_s/N^{1/3}$ The simplest form of correlation was used, namely, that based on Wigner's approximation formula

$$E_{corr} = \frac{-0.288}{R_s + 5.1} \quad . \tag{10}$$

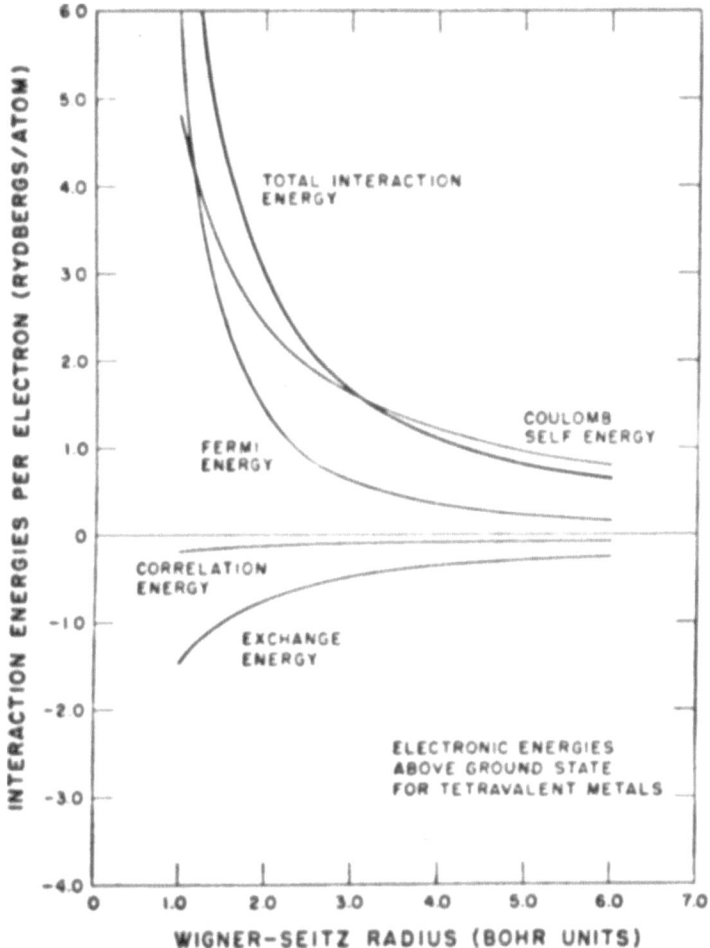

Fig. 2. Plot of the different electron-electron interaction terms
in Eqn. (9) (after Larson and Waber [7]).

    While improved treatments of exchange are available, such as
those by Slater and his colleagues [26] and by Liberman [27], significant precision should not be expected with this oversimplified
model. Typical interaction terms which are independent of the
element involved are presented in Fig. 2 for N = 4. While it might
be desirable to employ an improved treatment of correlation done by
Pines and Nozières [28], it is not likely to lead to significantly
different results. Just as Slater has indicated in his recent

studies that the exchange term should be multiplied by a constant
α_s, one could correct the kinetic energy by dividing by the effective
mass, m_e^*. In fact, each of the coefficients might be regarded as
adjustable parameters.

It has been decided to apply the more stringent test to the
model, namely, the original free-electron coefficients have been
used in this heuristic study.

The energies $E_0(R_s)$ and $E_T(R_s)$ are presented in Fig. 3 for Ti
and Zr. The equilibrium radius corresponding to a minimum in $E_T(R_s)$
is designated ρ. Even though the interaction energies depend only
on N and are independent of the specific metal being studied, the
curves of $E_0(R_s)$ may be different for two very similar metals such as
Ti and Zr.

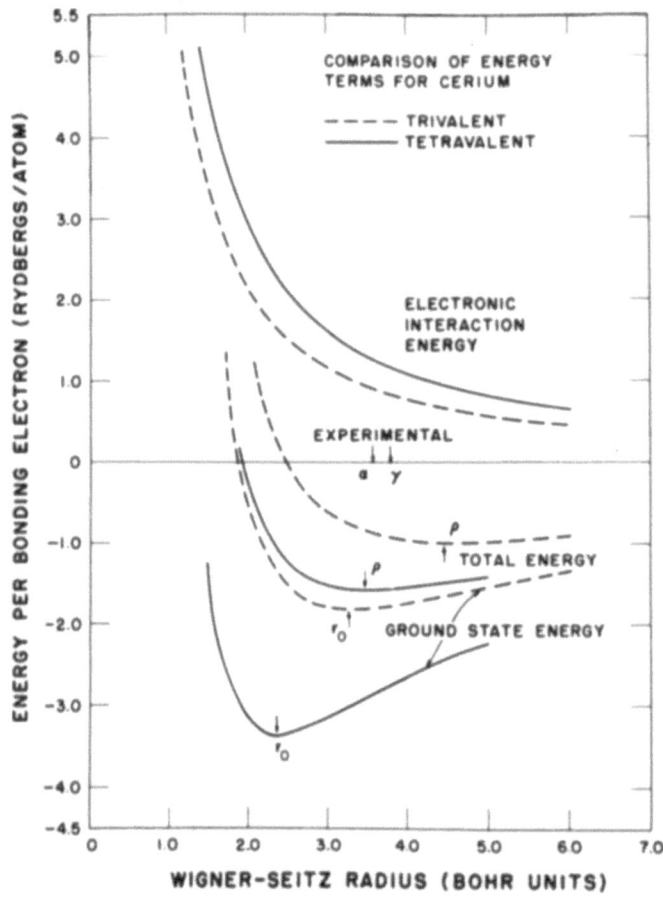

Fig. 3.   Determination of R_0 and ρ from E_0 and E_T curves of Ti and Zr.

2.1   Application of the Fröhlich-Raimes Model to the Estimation of Other Physical Quantities

Brooks [1] noted that the experimental cohesive energy per valence electron is surprisingly constant. For 23 elements, it ranges from 0.060 for Cs to 0.137 Rydbergs per atom for C.

The cohesive energy data tabulated for the lanthanides by Gschneidner [29] and the data of Trulson, Hudson, and Spedding [30] for the heat of sublimation at 298 K were divided by the number of valence electrons. This energy was found to vary for the lanthanides from 0.106 Rydberg for La to 0.0636 for Tm.

The cohesive energy S, i.e., the heat of sublimation at 0 K, can be obtained from the total energy at ρ by means of the expression,

$$S = \sum_{j}^{N} I_j - NE_T(\rho) \quad , \tag{11}$$

wherein the several ionization potentials I_j are summed.

The pressure-volume relation which leads to the compressibility χ at any pressure can be obtained in a straightforward manner from the $E_T(R_s)$ curve. Raimes [4] points out that the pressure P corresponding to a relative change in volume can be obtained from the expressions

$$P = \frac{-N}{4\pi R_s^2} \left(\frac{dE_T}{dR_s} \right) \tag{12}$$

$$-\frac{\Delta V}{V} = 1 - (R_s/\rho)^3 \quad . \tag{13}$$

Note that 1 atomic unit of pressure used in Eqn. (12) is equal to 150×10^6 kg/cm^2. No attempt has been made here to correct these data to absolute zero although, generally speaking, this change would tend to improve the agreement between calculated and experimental values. The compressibility K at zero pressure can be obtained from the second derivative of $E_T(R_s)$. With one exception, the listed values of the experimental compressibility were all taken from Bridgman's tabulation [31] of the quantity K using the more recent data for 75°C. The value for Si is from Pearson and Brattain [32].

Experimental and calculated values of the three physical constants ρ, S, and K are presented in Table I for a number of metals

Table I. Comparison of the Calculated Values of the Metallic Radius, ρ, Heat of Sublimation S and Compressibility K with the Experimental Values

Element Considered	Radius r_o, (Bohr Units)	Radius ρ*	Heat S (kcal/ mole)	Compressibility** $(10^6 \times K(cm^2/kg))$	Total E_T (kcal/mole)
Trivalent Elements					
Aluminum	2.130	3.188	23	0.775	1251
Exp.	--	2.990	78.1	1.34	1306
Scandium	2.540	3.635	98	1.262	1115
Exp.	--	3.427	89.9	--	1107
Gallium	1.929	2.971	10	0.597	1330
Exp.	--	3.161	64.8	2.0	1385
Yttrium	3.196	4.355	45.8	2.49	948
Exp.	--	3.760	100.8	2.95	1003
Indium	2.171	3.233	21	0.815	1236
Exp.	--	3.478	58.1	2.5	1273
Lanthanum	3.463	4.649	58	3.19	893
Exp.	--	3.921	103.1	3.24	938
Thallium	2.008	3.056	- 1.13	0.662	1298
Exp.	--	3.577	43.4	2.77	1341
Tetravalent Elements					
Silicon	1.844	2.874	-35	0.305	2344
Exp.	--	3.184	106.7	1.00	2486
Titanium	1.946	2.986	166	0.352	2267
Exp.	--	3.178	111.8	0.868	2213
Germanium	1.866	2.897	-41	0.315	2328
Exp.	--	3.308	88.8	1.364	2458
Zirconium	2.575	3.690	98	0.776	1881
Exp.	--	3.347	144.2	1.106	1927
Hafnium	2.670	3.797	25	0.864	1833
Exp.	--	3.301	148.4	0.881	1956
Thorium	3.092	4.273	83	1.351	1648
Exp.	--	3.754	142.9	1.846	1708
Pentavalent Elements					
Vanadium	1.627	2.620	97	0.141	3873
Exp.	--	2.817	122	0.609	3898
Niobium	2.237	3.310	55	0.336	3158
Exp.	--	3.071	175	0.57	3278

* The experimental values of the Wigner-Seitz radius were obtained from the volumes, V cm³, per gram atom listed by Teatum, Gschneidner, and Waber [33] using the formula $\rho = 1.3880\ V^{1/3}$.

**Experimental compressibilities are for 343 K, taken from the tabulation of Bridgman [31].

and semi-metals. Inasmuch as S is the difference of two large terms, the effect of any error in estimating the individual electronic energies is greatly magnified. To partially overcome this objection, an experimental value of E_T was obtained by adding the several experimental ionization potentials $\sum I_j$ to the experimental heat of sublimation. These two values of $E_T(\rho)$ are included in the table. The energy conversion factor used was 1 Rydberg/atom = 313.66 kcal/mole. The values of the heat of sublimation, S, for monatomic gaseous species corrected to 298 K were taken from Teatum et al. [33]. Whenever possible, the ionization potentials obtained by C. E. Moore-Sitterly [34] from spectroscopic data were used. One of the most readily usable and complete tabulations of ionization potentials has been presented by Finkelnburg and coworkers [35]. The ionization potentials for Hf III and Hf IV were recently given by Klinkenberg, van Kleef, and Noorman [36]. The ionization potentials for Th are not well known; the sum is probably accurate to better than 5%.

Values of K, S, and ρ were obtained independently by Raimes [37] who used the simplified theory cited above [4]. For comparison, these values for trivalent and tetravalent metals are presented in Table II.

In the calculation by Bernstein [6] of the cohesive energy of the lanthanides, the experimental values of the third ionization potential I_3 were used, where available, to obtain the R_0. Waber and Larson [38] recalculated ρ, K, and R_0 for a number of rare earth metals. These data are presented in Table III. Values are given for both divalent and trivalent forms of Yb and Eu. The use of both valences is much more questionable in the case of Lu.

Raimes [5] pointed out that, aside from the systematic errors due to the approximations made in deriving the equations, one source of error in the cohesive energy arises from the estimation of the ground state energy; a 2% change in $E_0(\rho)$ yields almost a 20% change in S. It is shown in Fig. 1 that for certain regions of the ionization potential, an appreciable error could be made in determining R_0 by means of the Raimes equation due to the steps. Such errors would have a stronger effect on the ground state energy at $R_s = R_0$ than at $R_s = \rho$. However, here we will not go into problems of estimating R_0 accurately.

Brooks [39] presented a simplified theory of cohesion which involved a modification of the Quantum Defect Method used by Kuhn [40] to determine $E_0(R_0)$. Values of r_0 determined in this way are compared in Table IV with those computed herein. The equilibrium radius ρ was evaluated using Eqns. (8) and (9) as above. The values of ρ obtained in these two ways are also compared with the experimental value in Table IV. It will be seen that Brooks' small r_0 values result in better agreement with the experimental radii. Also,

Table II. Unpublished Calculations by Raimes--
Metals of Higher Valency

Metal	Lattice Constant ρ (Å)		Compressibility $10^{12}K(cm^2/dyne)$		Cohesive Energy S(kcal/mole)		Deviation ΔS as % of Fermi Energy
	Calc.	Obs.	Calc.	Obs.	Calc.	Obs.	
Trivalent Metals							
Al	1.72	1.58	0.9	1.4	8	55	-12
Sc	1.99	(1.8)	1.5	---	75	70	2
Y	2.36	1.97	2.8	---	28	90	-29
Ga	1.60	1.67	0.7	2.1	- 6	52	-12
In	1.74	1.84	0.9	2.7	6	52	-11
Tl	1.65	1.89	0.7	2.3	-17	40	-13
Tetravalent Metals							
Ti	1.61	1.62	0.4	---	132	100	4
Zr	2.02	1.71	0.9	---	14	110	-20
Sn	1.74	1.86	0.5	1.9	-36	78	-18
V	1.42	1.48	0.2	---	100	85	+ 1
Semiconductors							
Si	1.55	1.68	0.3	0.3	-63	85	-18
Ge	1.53	1.74	0.3	1.4	-59	85	-17

* The author is indebted to Prof. Raimes who made this information
available for use.

Table III. Comparison of Results Obtained in Recent Cohesive Energy
Calculations for the Lanthanide Metals

	Wigner-Seitz Radius, ρ (Bohr units)			Compressibility 10^6 x K(cm^2/kg)			Parameter, R_o (Bohr units)	
	BS	LW	OBS.	BS	LW	OBS.	LW	BS
Sc(3)	3.69	3.635	3.427	1.37	1.27	--	2.54	2.56
Y (3)	4.40	4.358	3.760	2.69	2.44	2.95	3.198	3.20
La(3)	4.70	4.653	3.921	3.47	3.22	4.13	3.466	3.47
Ce(3)	4.65	4.464	3.81 (γ)	3.29	2.75	4.10(γ)	3.295	3.42
Ce(4)	4.31	3.490	3.57 (α)	1.58	0.61	4.72(α)	2.387	3.09
Sm(3)	4.50	4.174	3.761	2.88	2.15	3.52	3.684	3.28
Eu(2)	4.84	4.829	4.290	8.13	7.76	4.91	3.715	3.67
Eu(3)	--	4.088	4.290	--	1.98	4.91	2.953	--
Yb(2)	4.54	4.485	4.044	6.33	5.92	7.84	3.389	3.28
Yb(3)	--	3.986	3.986	--	1.80	7.84	2.860	--
Lu(3)	4.25	3.771	3.628	2.36	2.51	2.33	3.198	3.06
Lu(2)	--	4.358	3.628	--	3.10	2.33	2.701	--

BS = Ref. [6]; LW = Ref. [38]

Table IV. Comparison of the Brooks' Values of R_o and ρ with Present
Computations of these Radii in Bohr Units .

Element Treated	Ground State Radius, R_o		Wigner-Seitz Radius, ρ		
	Brooks	Present	Present	Brooks	Exper.
Na	2.97	3.244	4.147	3.886	3.991
Mg	2.38	2.563	3.601	3.437	3.345
Al	2.03	2.132	3.190	3.088	2.990
Si	1.755	1.846	2.875	2.75	3.184
Zn	--	2.003	3.024	2.784	2.904

the experimental values of S for the five metals studied by Brooks
agree with his calculated values better than they do with those
calculated by the present method.

The Wigner-Seitz radius ρ and cohesive energies of the noble
metals Cu, Ag, and Au are compared with experimental values in
Table V. Brooks [1] noted that the experimental values of S for
these monovalent metals were three or four times larger than the
values of S/N for any of the other 32 elements he considered. He
concluded that the closed d-shells contribute significantly to the
binding. It is interesting that for the IIB, IIIB, and IVB elements
which follow these transition metals in the Periodic Table, the
binding is normal and the values of S/N lie in the same range as
mentioned in the first part of this paper. Thus, the ion cores are
more effectively separated from each other by the gas of free electrons; an error arises when one assumes that the d-electrons are
confined to a very small region associated with the ion core and
have negligible probability of being found at the cell boundary.
This is equivalent to the old tight binding approximation which is
inadequate for transition metals.

Table V. Comparison of Calculated and Experimental Values for the
Noble Metals (after Brooks [1])

	Cohesive Energy, S Rydbergs/electron		Wigner-Seitz Radius (Bohr units)		Compressibility 10^6 x $K(cm^2/kg)$	
	Calc.	Exp.	Calc.	Exp.	Calc.	Exp.
Cu	0.087	0.260	2.608	2.669	2.80	0.734
Ag	0.086	0.221	2.666	3.017	3.03	1.004
Au	0.085	0.264	2.146	3.012	1.43	0.570

Typical behavior of the calculated compressibility K as a function of pressure is shown in Fig. 4 for four tetravalent metals.

In the noble metals, the interatomic distance, however, is
governed by the interaction between d-electrons and the ion cores.
Contrary to the trend found for most metals in Table I, the calculated values of ρ in Table V are less than the observed values. Thus,
one would anticipate that, due to ionic repulsion, the actual compressibilities of Cu, Ag, and Au would be substantially lower than
those calculated by the present method, which only includes electron-
electron repulsions. This is borne out in Table V where the experimental K values are only 0.3 to 0.4 those of the calculated values.

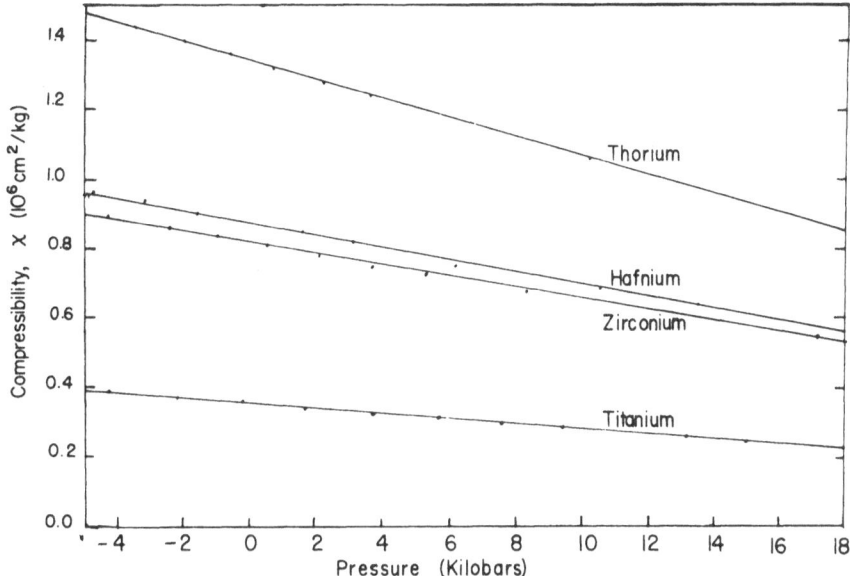

Fig. 4. The calculated pressure dependence of K for four different tetravalent metals.

Thus, it is reasonable to conclude that for these noble metals, the interaction between the valence electrons and those forming the ionic core is much stronger than in normal metals and that the present simple method is not satisfactory.

The importance of such repulsion is illustrated in the pressure-volume graph, Fig. 5, for Ni. As a shock wave passes through a metal, it is subjected to large negative pressures during the rarefaction wave--although that portion is not plotted in Fig. 5. As an atom, Ni has only one 4s electron to act as a valence electron. To incorporate the effect of the 3d electrons, we have assumed that N is higher than 2. In doing the calculations, N = 3 gives better agreement with the experimental shock wave data [41].

A more significant way [7] of dealing with the ion cores is to add

$$E_{ionic} = b \exp\left(\frac{-S_{ij}}{\rho_{BM}}\right)$$
(14)

to the terms in Eqn. (9).

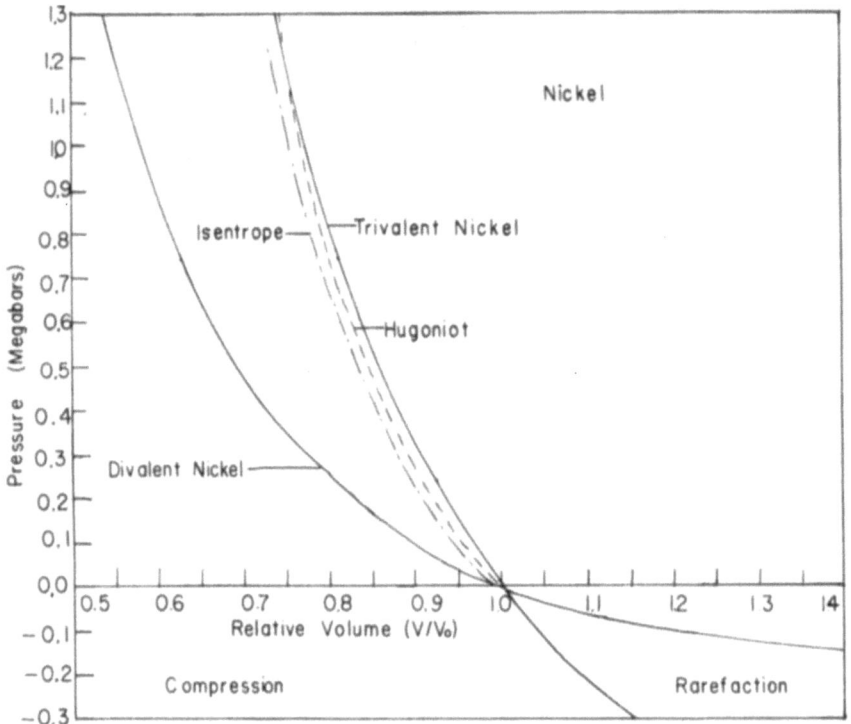

Fig. 5. Comparison of experimental shock wave data of McQueen and
Marsh [41] for Ni with two calculated P versus V curves. The higher
valence N = 3 (which is chemically stable) is an approximate way of
taking account of repulsion between d-electrons and ion-cores.

Here, b is an adjustable constant approximately equal to 10^{12}
ergs, and S_{ij} is given by

$$S_{ij} = r_{ij} - r_i - r_j \quad , \tag{15}$$

where r_i and r_j are the ionic radii such as most chemists use. Concerning the repulsive parameter ρ_{BM} which occurs in the Born-Mayer
expression for ionic crystals, Hafemeister and Zahrt [42] showed
that values of ρ_{BM} could be calculated directly from overlap integrals; normally, it is an experimentally adjustable quantity approximately equal to 0.345 Å. We have established that inclusion of ionic
terms improves the agreement with the shock wave data on Au, as Fig.
6 shows. Inasmuch as the inclusion of ionic repulsion has been made

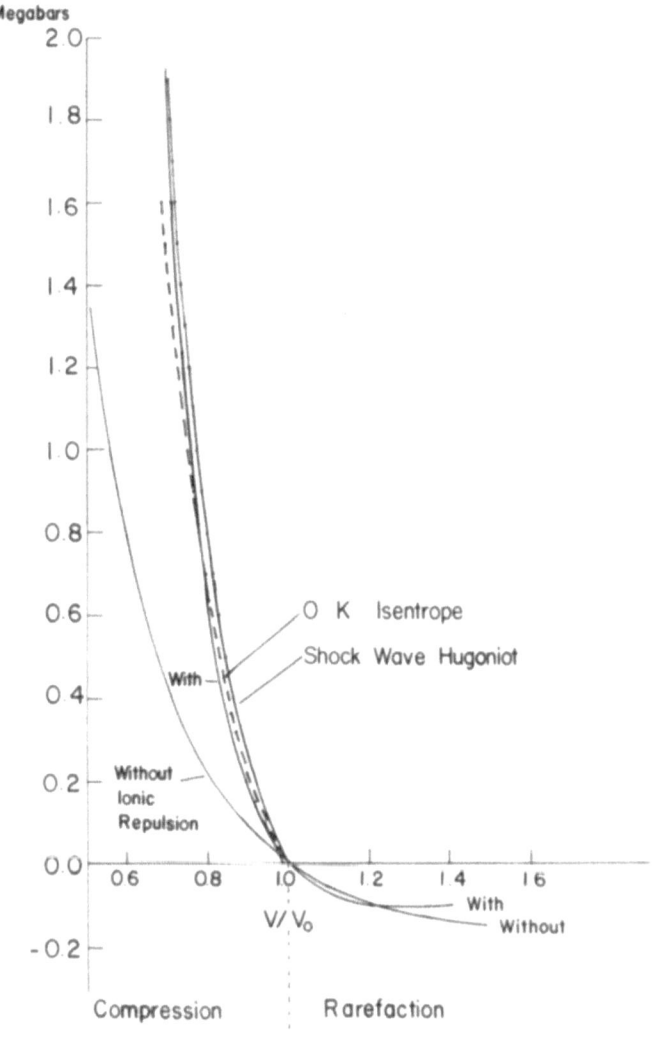

Fig. 6. Similar shock-wave data for Au compared with those calculated from $E_T(R_s)$ with and without E_{ionic} from Eqn. (14) included.

only for limited cases here, it would take us away from the main theme of this paper to go into further detail about how the quantities b and ρ_{BM} are estimated. This will be discussed in a separate paper.

2.2 Application of the Fröhlich-Raimes Model to Alloys

The Fröhlich-Raimes method has been applied to alloys [38]. One reasonable assumption is that the ground state energy of the alloy is given by a linear combination of E_o terms. That is,

$$-E_o \; (C_A, R_S) = \frac{C_A n_A}{R_s} \left(3 - \frac{R_s^2}{R_{oA}^2} \right) + \frac{C_B n_B}{R_s} \left(3 - \frac{R_s^2}{R_{oB}^2} \right) \quad , \qquad (16)$$

where C_A, n_A, and R_{oA} are, respectively, the atomic concentration, the valence, and the Fröhlich-Raimes parameter for component A, and R_S is the variable radius of a Wigner-Seitz sphere. Here it is assumed that $\ell = 0$. For the interaction energy, the mean number $C_A n_A + C_B n_B$ of valence electrons was used. This mean value, $\bar{n}$, is inserted in the remaining terms of Eqn. (9) rather than averaging the two E_t values. To distinguish expressions, we will denote the total energy of an alloy by E_t. Values of E_t, radius ρ, and compressibility χ have been obtained as a function of concentration for a large number of alloys. Most of these results will be reported elsewhere, but one type of result will be summarized here.

It has become common for metallurgists to compare the observed average lattice spacings of alloys with those obtained by averaging the lattice spacings of the components. One can calculate $\rho(C_A)$ and compare it with Vegard's Law which states that the volumes of components are additive. We have assessed this deviation from Vegard's Law by computing

$$-y(C_A) = \rho(C_A) - (C_A \rho_A + C_B \rho_B) \quad . \qquad (17)$$

Such deviations, $\Delta\rho(C_A)$ are illustrated in Fig. 7 for Ti and two different valences of Ce. A large negative deviation can be seen for the tetravalent Ce.

The total energy $E_t \; (C_A, \rho)$ has been plotted for three other binary systems. Fig. 8 is for Na-Ag, for which little solubility exists at either end. However, the calculated curve cannot illustrate this feature. The large positive deviation from Vegard's Law which can be seen in the plot of $\rho(C_A)$ is the clue. Of course, strain energy in the lattice has not been included in the present model, so direct evidence for phase separation cannot come from the intentionally restricted model used at the present time. Fig. 9 is for the Na-Mg alloys. The heat of formation of these alloys, ΔE, is the deviation from the straight line which would be given by a mechanical mixture. In almost all cases studied, the solid solutions would tend to be more stable than a two-phase mixture. Note that this stability does not result from including the usual entropy of

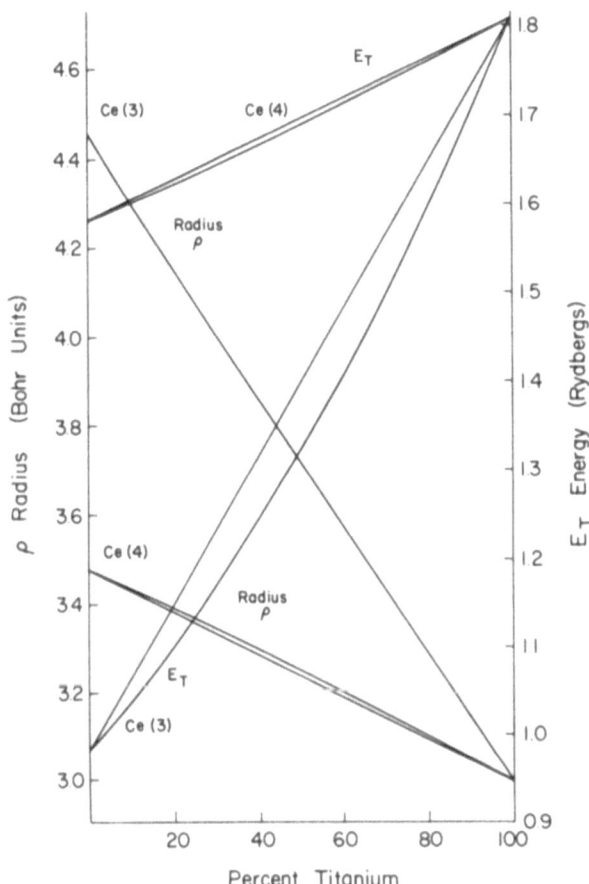

Fig. 7. Plot of $-E_t(C_A,\rho)$ evaluated at the composition–dependent radius $\rho(C_A)$ for Ce–Ti alloys. The effect of assuming that Ce is tetravalent is to cause only a slight depression in the E_T curve. It is interesting that $\rho(C_A)$ lies above the straight line connecting the ρ values for the two pure materials. A positive deviation is indicated from Vegard's Law of additive atomic radii.

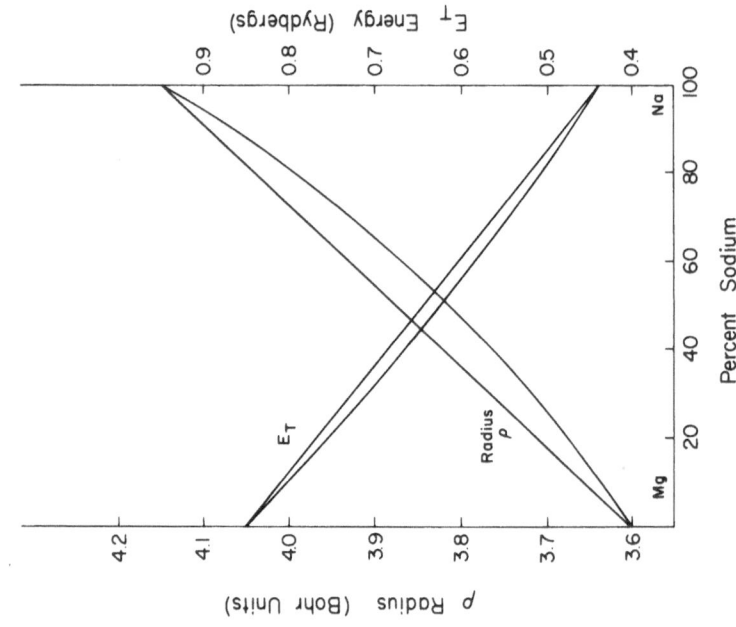

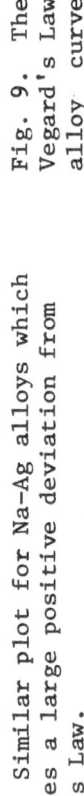

Fig. 9. The strong negative deviation from
Vegard's Law for Mg-Na alloys contrasts with the
alloy curves presented so far.

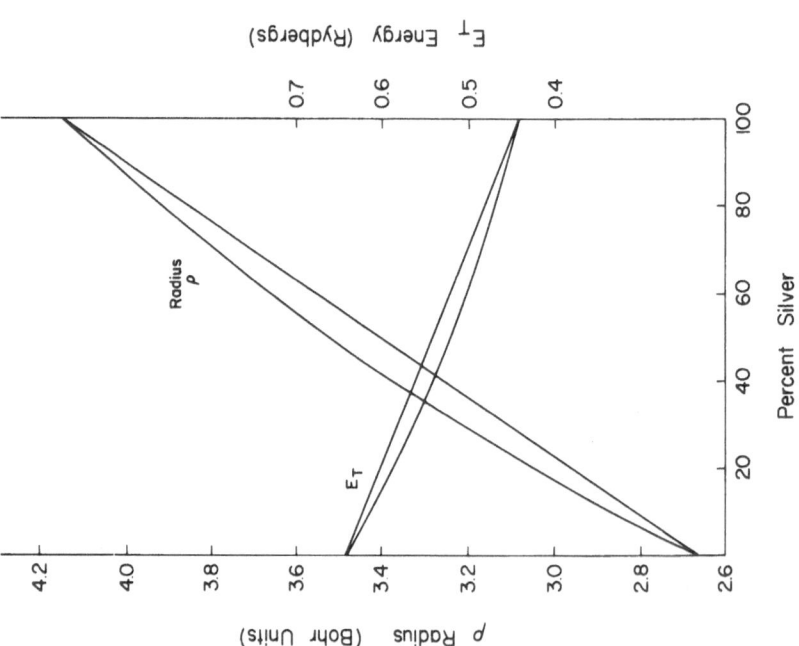

Fig. 8. Similar plot for Na-Ag alloys which
indicates a large positive deviation from
Vegard's Law.

mixing, but is strictly electronic in origin. No electronic entropy term has even been included. Both would increase the stability of any solid solution.

The results of a large number of calculations about Vegard's Law are summarized in Table VI. In some cases, there is experimental evidence for positive and negative deviations. In a number of cases, predictions are made which may be verified using additional lattice parameter data which were not readily available to the author when this summary was originally made.

The calculated data in Table VI are arranged by the valence of the solvent. The cases treated were for binary systems cited by Pearson [43] in his large tabulation of lattice parameters. Without becoming involved in detailed evaluation of the experimental deviations of different alloys, or decisions regarding the comparative reliability of specific experimental data, it appeared to be sufficient at this time to report the sign of the deviation. One complication, for example for the Mg-based alloys, is that small alloy additions do not stabilize the ε cubic phase. Since the c/a ratio of Mg is not ideal, packing of individual spherical cells in which the electrons exert a pressure can not be the sole criterion for determining the volume. Some modification of the model to include either structure-sensitive properties or perhaps covalent bonding may be necessary to improve our understanding of the lattice dimensions in these alloys.

One illuminating test of this simple electron gas theory is the group IVB metals such as Ti, Zr, and Hf, which do exist in both cubic and hexagonal modifications. In the upper portion of this drawing, the a_0 is plotted for the cubic phase Zr-Ti alloys. However, in the middle portion, two separate branches are shown for the a_0 parameter of hexagonal Zr-based and Ti-based alloys. The associated c_0 values are presented in the lower portion of Fig. 10. These might be interpreted in terms of the definitely positive deviation from Vegard's Law presented in Fig. 11. Phase separation to the larger cubic phase would be compatible with the present model. At this time, it is not clear why Hf-Nb alloys remain cubic nor why the Zr-Hf alloys do not form a cubic β phase.

The relative deviations $100y(c_A)/\rho_A$ can be used to reduce confusion about units, and values are presented in Table VII for Ce-based alloys containing 10% of several solutes. The relative deviations have the correct sign in both Th-poor solid solutions when compared with the data of Waber, Harris, and Raynor [44], but are smaller in magnitude than the observed values (4.14 for α and -1.31 for γ). The relative deviations they computed using Friedel's Elastic Model [45,46] are similar (3.14 for α and -0.57 for γ) to those listed above.

Table VI. Calculated Deviations from Additive Volumes Compared
with Experimental Data

Solvent-Solute	Solute Conc. Selected for Calculation of $\Delta\rho^3/<\rho^3>$	$\dfrac{100\Delta\rho\,(a)}{<\rho>}$	$\dfrac{100\Delta\rho^3}{<\rho^3>}$		Remarks
			Calc.	Obs.	
Monovalent Solvent					
Li–Mg	50	neg.	−3.6	−8.0	
Na–Mg	50	neg.	−5.9		Pre-1910 work
Na–Zn	50	neg.	−13.8		Pre-1910 work
Na–Ag	50	pos.	−3.9		
K–Rb	50	pos.	−0.1		Miscible
K–Cs	50	pos.	−0.4	−2.9	
Rb–Cs	30	pos.	−0.2	−3.5	
Cu–Ag	50	pos.	−0.01	+0.9	Quenched
Cu–Au	50	pos.	−0.6	+0.3	Metastable
Ag–Au	50	pos.	−0.8	−0.5	
Ag–Cd	9.56	pos.	+3.6	−1.1	
Cu–Zn	33.56	pos.	+5.7	−2.4	
Divalent Solvent					
Cd–Ag	5	pos.	+0.8	−0.6	
Mg–Li	14.01	neg.	−1.6	−3.9	
Mg–Al	22.4	neg.	−2.5	−1.3	
Mg–Zn	1.2	pos.	−0.03	−0.02	
Mg–In	7.23	neg.	−0.9	+0.9	
Mg–La	1.5	pos.	+0.45	−0.1	
Mg–Sn(2)	2.58	pos.	.0	−0.5	Based on β-Sn
Mg–Sn(4)	2.58	neg.	−0.7	−1.3	Based on α-Sn
Pb(2)–Tl	50.4	neg.	−4.8	+0.4	
Sn(2)–Sb	8.1	neg.	−4.6	−1.8	
Ca–Ba	50	pos.	−0.9	0.5	
Sr–Ba	50	pos.	−0.3	−0.8	
Zn–Cu	1.85	pos.	+0.3	−1.2	
Sn(2)–In	5	neg.	−0.8	+0.03	

Table VI. (Continued) Calculated Deviations from Additive Volumes
Compared with Experimental Data

Solvent-Solute	Solute Conc. Selected for Calculation of $\Delta\rho^3/<\rho^3>$	$\frac{100\Delta\rho\,(a)}{<\rho>}$	$\frac{100\Delta\rho^3}{<\rho^3>}$ Calc.	Obs.	Remarks
		Trivalent Solvent			
Al–Zn	30	pos.	+1.0	+0.8	
Al–Cr	1.24	neg.	−0.8	−0.03	
Al–Mg	35.8	neg.	−3.2	−1.7	
Al–Ti	0.2	neg.	−0.01	−0.14	
In–Cd	4.02	pos.	+0.03	1.0	
In–Pb(2) }	13.07	{neg.	−1.4}	0.5	
In–Pb(4) }		neg.	−0.6}		
In–Sn(2) }	9.39	{neg.	−1.4	−2.4	Based on β–Sn
In–Sn(4) }		neg.	−0.2	+0.2	Based on α–Sn
In–Tl	19.4	pos.	−0.04	+0.2	
Bi–Sb	48.2			+0.3	
		Tetravalent Solvent			
Ti–Al	34.01	neg.	−1.2	−3.0	
Ti–Cr	16	neg.	−3.9	−0.3	
Ti–Hf	60	pos.	−1.4	+0.9	
Ti–Nb	54.5	pos.	+1.3	−0.9	
Ti–Mo	23.27	pos.	+0.3	−2.6	
Ti–Sn(2) }	8.5	{neg.	−3.4	−6.2	Based on β–Sn
Ti–Sn(4) }		pos.	−0.03	−3.2	Based on α–Sn
Ti–V	50	neg.	−2.4	−0.1	
Ti–Zr	30	pos.	−1.2	−0.8	2-phase region
Zr–Ti	30	pos.	−0.9	−0.9	
Hf–Zr	38.2	pos.	−0.02	+0.3	
Zr–Th	49.1	pos.	−0.6	−6.5	
Zr–Nb	50	neg.	−2.0	−0.9	
Hf–Nb	.475	neg.	−2.6	−3.2	
		Pentavalent Solvent			
Cr–Mn	42.7	{neg.	−9.0 }	−1.5 {	Mn II
		pos.	+0.3 }		Mn V
		neg.	−0.3 }		Mn VII
Cr–Mo	50	pos.	−1.6	+1.8	
Cr–V	50	neg.	−1.6	−1.0	

Note: The linear combination of volumes $C_A\rho^3(C_A) + C_B\rho^3(C_B)$ does not
lead to the same deviations (occasionally even differing in sign) as
does the linear combination of cell diameters as used in Eqn. (17).

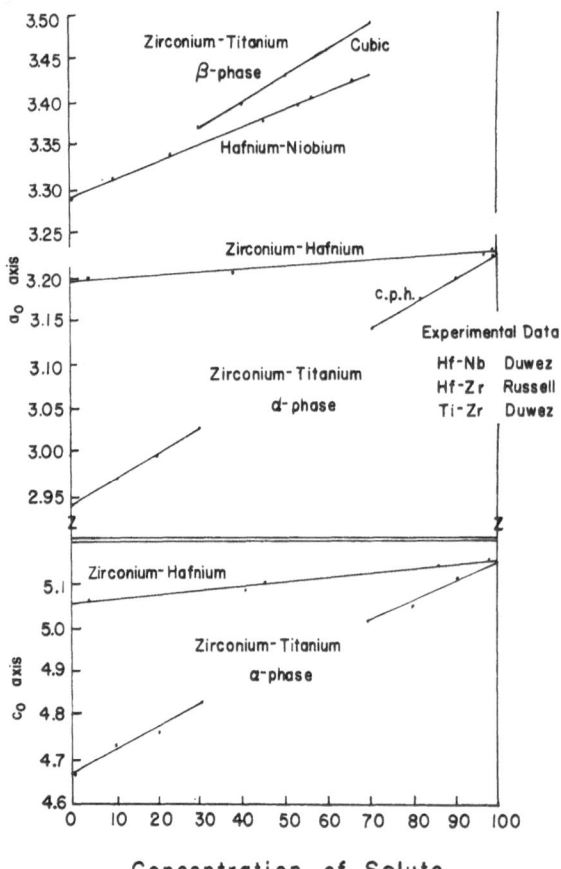

Fig. 10. Experimental values of the two lattice parameters of a series of hexagonal metals. In the middle of the composition range, Zr-Ti alloys become bcc at room temperature. Data were taken from Duwez, P., J. Appl. Phys. 22 1174 (1951), and Russell, R.B., J. Appl. Phys. 24 232 (1953).

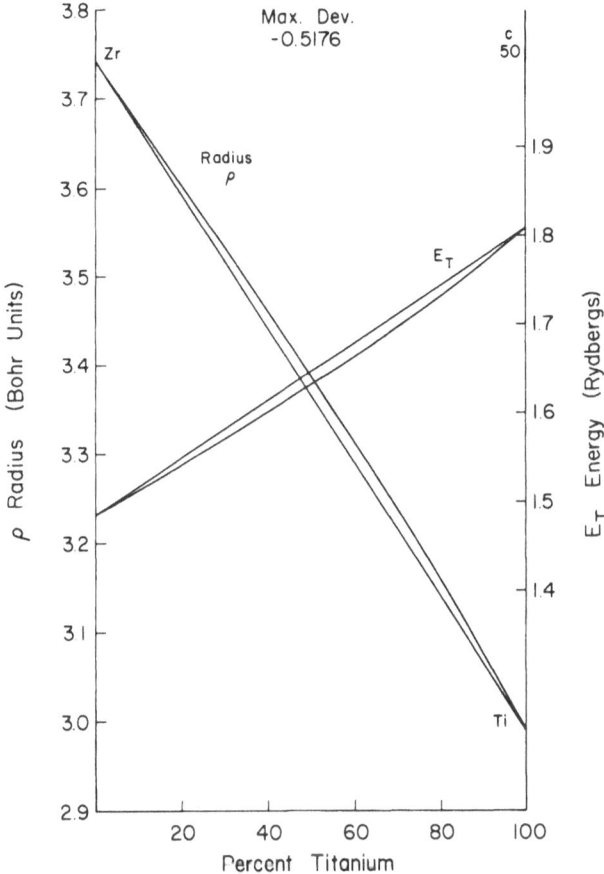

Fig. 11. Plot of $-E_t(C_A)$ and $\rho(C_A)$ for Zr-Ti alloys.

Table VII. Relative Deviations from Vegard's Law Calculated for Ce-Base Alloys

Solute	αCe	γCe
Mg	−1.47	+6.75
Al	2.07	+4.57
La	−1.84	−0.09
Zr	+0.32	−2.29
Th	+2.73	−0.89

The sign of $y(C_A)$ does agree with that reported by Gschneidner for dilute Mg alloys, but does not agree well with the findings of Harris and Raynor [47] for Ce dissolved in α-Zr.

Davison and Smith [48] estimated the enthalpy of formation of $CaMg_2$ in a relatively similar manner. They indirectly allowed for the shift in the bottom of the band $E_o(R_s)$ with concentration. The volume of the intermetallic compound was found from the equation

$$\frac{4\pi}{3} R^3 = \frac{4\pi}{3} (C_A R_{SA}^3 + C_B R_{SB}^3) \quad . \tag{18}$$

Further, they assumed that R_{SA}/R_{SB} was 1.225 to correspond to a Laves phase. In this author's opinion, it would have been preferable to have estimated the average number of electrons per atom and the calculated ρ and E_t values. They used Brooks' values of R_o for Ca and Mg. Their data are presented in Table VIII. Smith [49] has found this prescription works reasonably for a number of intermetallic compounds.

Table VIII. Comparison of Three Physical Quantities for Laves Phases Calculated by the Fröhlich-Raimes Method with Experimental Values (after Davison and Smith, Ref. [48])

	Heat of Formation $-\Delta H_f$ (kcal)			Volume of Formation $-\Delta V$ (Å^3)		Compressibility K(cm^2/kg)	
	Smith	Brooks	Exp.	Calc.	Exp.	Calc.	Exp.
$CaMg_2$	9.8	11.2	3.2	5.66	5.66	265	3.77
$SrMg_2$	7.9	9.39	1.7	9.64	8.42	--	--
$BaMg_2$	2.8	5.69	0.5	15.35	7.03	--	--

One interpretation of the failure of the Hf-Zr to separate into a cubic phase can be made in terms of the compressibilities of the host lattices. As Fig. 4 shows, the K values for Hf and Zr are very similar. In contrast, Ti is significantly less compressible. The present adaptation of the Fröhlich-Raimes model to binary alloys permits one to calculate the variation of K with composition. A set of curves for Ti-Zr alloys is shown in Fig. 12. This results from evaluation of Eqn. (16) and not from a simple scaling of the two P-V curves for the pure metals. We have not verified that the experimental compressibility of alloys is as variable as indicated by Fig. 12. The literature indicates that very little data have been

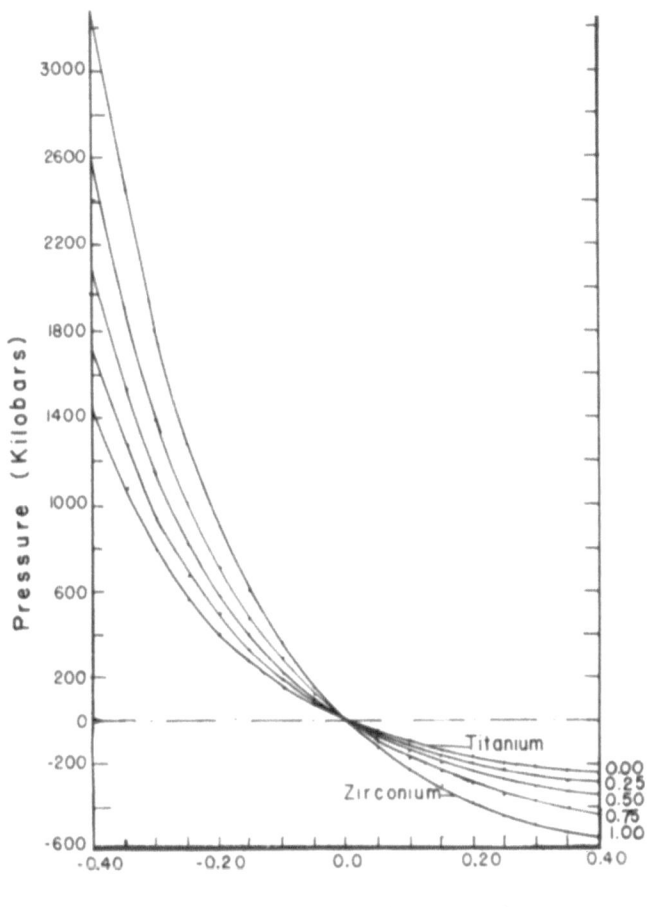

Fig. 12. Variation of the pressure volume curves calculated for Ti-Zr alloys.

collected. It seems that this would be a useful test which could be made of this volume-dependence theory.

To summarize, the major features of this FR model for a metal and its alloys depend on the relationship between the pressure and the density of electrons. This in turn depends on the volume in which they are confined. However, any details of the spatial arrangement of the atoms are ignored. As such, it leads to a good interatomic force law. In the Fröhlich-Raimes model, it is assumed that the lowest energy of an electron $E_T(C_A)$ lies between E_T for

one element and E_T for the other. Because of the simplicity of the
FR model, any Brillouin zone effects have not been dealt with.

2.3 Electron Cell Model

In preparing this manuscript, the author came upon a very
similar recent treatment of alloys by Bolsaitis and coworkers [50,
51,52]. It, too, is based on Fröhlich's equations [1]. However,
the spirit of their study is different in that they have paramater-
ized all terms and included several further thermodynamic correc-
tions. They represent the cohesive energy by the equation

$$S = I_1 + \frac{A}{R_s^3} + \frac{F}{R_s^2} + \frac{B}{R_s} + E_{corr} + 6E_p + E_{vib} \quad . \tag{19}$$

Coefficient A could have been obtained from Eqn. (3) and coefficient
F could have been obtained by combining the remaining terms in Eqn.
(3) with the third term in Eqn. (9). Since they dealt only with Cu,
Ag, and Au, B would have been equal to (1.2 - 0.916). Their E_p is
given by Eqn. (14) except that the r_i and r_j were not included in
the numerator. They also included a van der Waals attractive term
C/R_s^6, and a vibrational energy term. By virtue of having a considerable number of parameters to adjust, excellent agreement with the
experimental data can be obtained. They introduced further arbi-
trariness when dealing with alloys by integrating the Grüneisen
equation to get an entropy correction

$$S' = \int_\infty^T \frac{C_p - C_r}{T} \, dT = \int_\infty^T \alpha^2 VB_T dT \quad . \tag{20}$$

They further limited the ionic repulsive term from Eqn. (14) to
unlike pairs of atoms.

Bolsaitis and coworkers varied the local charge density, ρ,
cell radii according to Sanderson's [53] stability ratios SR, and
in later papers, they used a definition given by Coulson for the
stability. Both definitions assumed that the charge densities were
uniform within an "electron cell" and could be scaled. Specifically,
they chose the radii of the Cu and Au atoms following the prescription

$$\frac{\rho_{Cu}(r)}{\rho_{Au}(r)} = \frac{r_{Au}^3}{r_{Cu}^3} = \frac{SR_{Cu}}{SR_{Au}} \quad , \tag{21}$$

and the mean alloy volume according to

$$V = \frac{4\pi}{3} (\chi_{Cu} \, r^{\circ}{}_{Cu}{}^{3} + \chi_{Au} \, r^{\circ}{}_{Au}{}^{3}) \quad . \tag{22}$$

Since neither radius is its equilibrium value, they estimate the volume of mixing V_M by replacing r_{Au} and r_{Cu} by their equilibrium values $r_{Au}{}^{\circ}$ and $r_{Cu}{}^{\circ}$ in the parentheses and then multiplying it by $(1 + V_M)$. This is an <u>ad hoc</u> but plausible procedure.

Let us turn now to a second simple model which depends on knowing the band structure of each component. This model is sensitive to the details of the crystal arrangement. In general, it is available for only one interatomic distance and a complicated calculation is needed for each atomic volume as subtle changes may occur in the relation of one band or E(k) curve to another. For practical reasons, the following model may be regarded as independent of the atomic volume.

3. LINEAR COMBINATION OF DENSITY OF STATES

In the basic theory of band structures, one allows one state and two spins for each atom in the piece of metal, and in general, the discrete spectrum of E(k) is replaced by a continuum if 10^{23} atoms are involved. One observes, when one discusses impurity effects in semiconductors, that the interesting case occurs when the impurity level lies in the gap between the conduction and the valence bands. Since the individual impurities are far apart in general, and are screened by a dielectric medium between them, the levels can be treated as discrete. However, if several types of defects are present, several very narrow energy ranges, and hence spikes, appear in overall density of states. The heights of these reflect to a reasonable extent how many impurity atoms of a given type are present. As a second observation, where band structure calculations for intermetallic compounds have been made, one can frequently assign energy ranges where the states in the E(k) curves can be identified as coming primarily from states derived from one constituent. Examples of this are the important studies on β-CuZn alloys and isomorphic phases by Johnson and his colleagues [54,55]. The third observation comes from Soven's study of the Coherent Potential Model [56], for an AB alloy. This work shows a modest increase in N(E) values near ends of the allowed range of energies for constituent A and constituent B where $N_A(E)$ and $N_B(E)$ do not overlap -- otherwise, the N(E) curves appear additive.

These observations led Koskimaki and Waber [18] to propose as a working hypothesis that a simple linear combination of the two density of states would suffice to substantially model the case of randomly distributed solid solutions.

In order to do this adequately, it is necessary to abandon the traditional idea of measuring all states from the bottom of the energy range (usually a point in k-space called Γ). Both must be referred to a common energy. When the energies are measured with respect to vacuum, the E_F values for different metals, they occur in a relatively small range. This important observation has been discussed by Snow and Waber [14], whereas the bottom of each band falls (relatively) smoothly with the atomic number. The curves for the two pure metals are shown in Fig. 13.

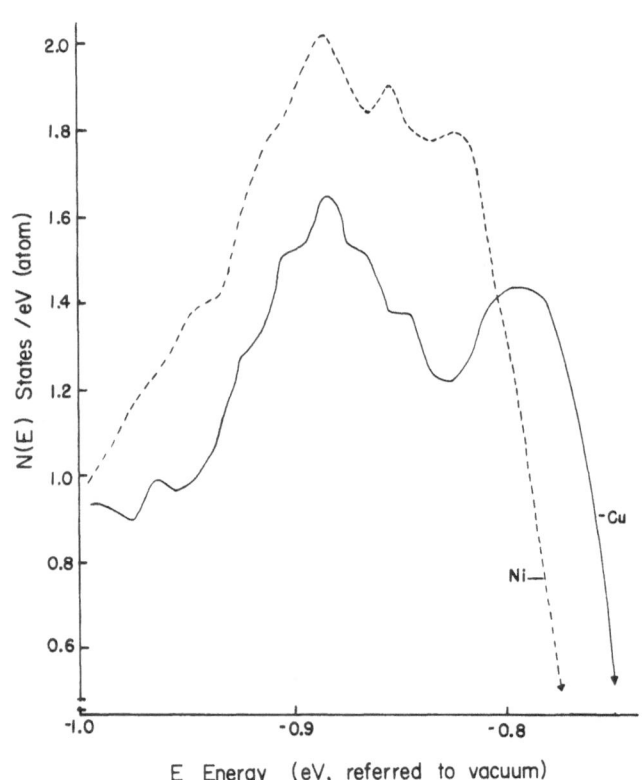

Fig. 13. N(E) curves for pure Cu and Ni obtained by Waber and Snow [14] using Slater's exchange parameter $X_\alpha = 1$. These two curves are combined to obtain Fig. 14.

   The LCDS model gives answers readily if the N(E) data are
available for both constituents, and these alloy N(E) curves are
in good agreement with the experimental facts. In Fig. 14, a typical calculation of the composition-dependent $N(E^a)$ curves are presented for a solid solution of Cu and Ni. It is very interesting
that the set of these curves obtained with the LCDS model shows
many of the features exhibited in alloy results of Eastman [57]
and Spicer [58] who used photoemission spectroscopy; and Clift,
Curry, and Thompson [59] who used soft X-ray spectroscopy. At
least the morphology of the alloy N(E) curve is compatible with
experience. No detailed comparison is appropriate at this stage of
development.

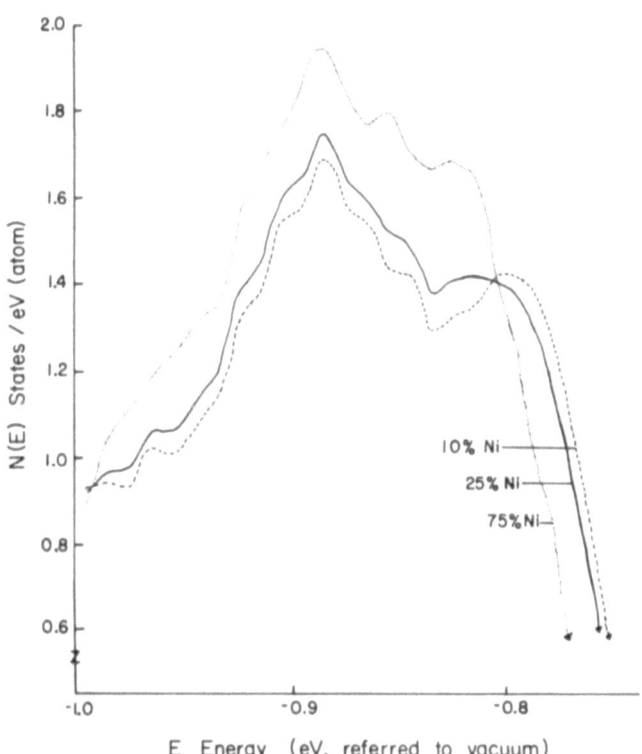

Fig. 14. Set of density of states $N(E^a)$ curves calculated for
several Ni-Cu alloys using the LCDS model of Koskimaki and Waber
[18].

One observation might be made in passing. Since the individual
values of E_F are similar in energy, no large transfer of electrons
to one type of atom occurs. Charging effects such as those discussed at the recent conference [60] will be less than if one combined the N(E) curves after the bottoms of the two bands had been
aligned at the arbitrary zero.

Now to locate the Fermi level for the alloy, one knows that
the number of electrons is $C_A n_A + C_B n_B$ if there are no major charge
transfer effects. The bottom of the alloy band will occur at the
lower energy of the two N(E) curves although it may not be heavily
weighted. One finds the mean number of valence electrons $\bar{n}$ per atom
from $C_A n_A + C_B n_B$. Then one fills up the weighted states of the
linearly combined $N_A(E)$ and $N_B(E)$ data to contain $\bar{n}$ electrons. Thus,
the E_F^a values will have a definite composition dependence, as will
$N(E_F^a)$. $N(E_F^a)$ curves obtained in this manner for Ni-Cu alloys are
presented in Fig. 15.

For background information, the LCDS curves of $N(E_F^a)$ are presented in Fig. 15 for Ni and Cu, and in Fig. 16 for Ni-Al alloys.
The decrease in $N(E_F^a)$ near 60% solute has been attributed to filling

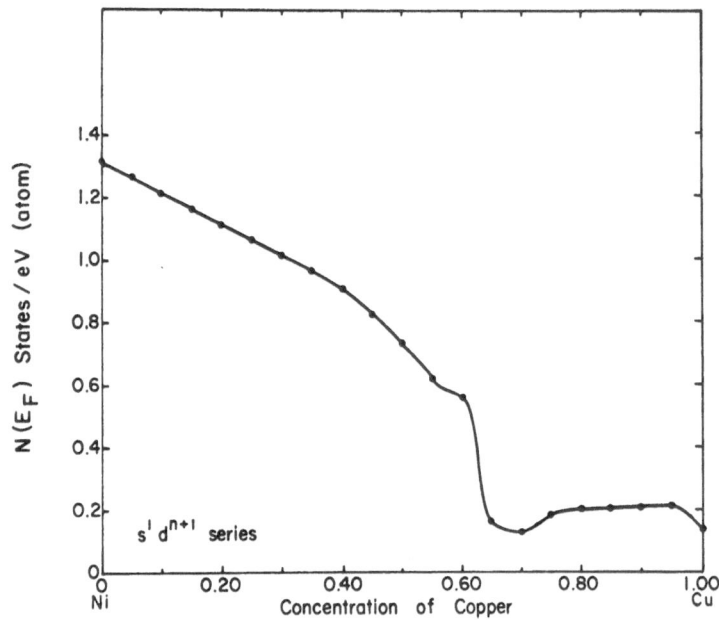

Fig. 15.   Calculated density of states at the Fermi level $N(E_F^a)$ for
homogeneous Ni-Cu alloys. The individual Fermi levels were not shown
in Fig.14.

of the d-states in Ni. Analysis of the data shows that some d-like
states are unoccupied even when the concentration of solute exceeds
the traditional amount needed at room temperature to fill 0.6 hole
in Ni.

Examples of these $N(E_F^a)$ curves are presented for several binary
alloys below. The curve for Ti plus Al in Fig. 17 is rather unin-
teresting since no sharp peaks and deep valleys occur in either of
the component $N(E)$ curves.

The case of Ti plus V, which is shown in Fig. 18, is more
interesting, particularly at the V-rich end. Two curves have been
shown in Fig. 19 for the solution of Ti with Mo. They are based
on two different configurations of Ti, used as input to the APW
program. It is more likely that the effective configuration in the
solid bcc phase is d^3s^1, as Snow and Waber [14] discuss. Despite
whichever curve one chooses to use, the shape of the $N(E_F^a)$ curve
is surprising since one might at first have anticipated simple curves

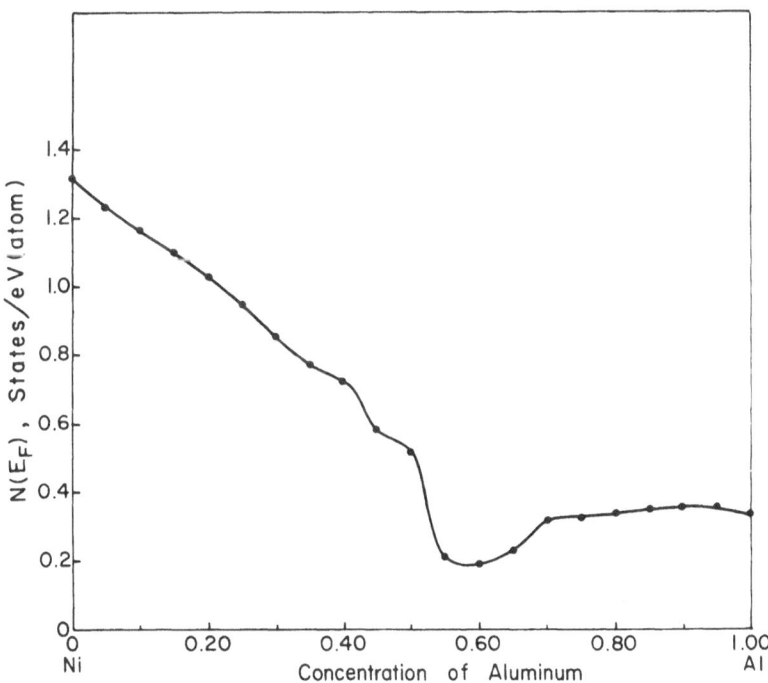

Fig. 16. Composition-dependent change in the density of states
$N(E_F^a)$ at the calculated Fermi level E_F^a for homogeneous Ni-Al alloys.

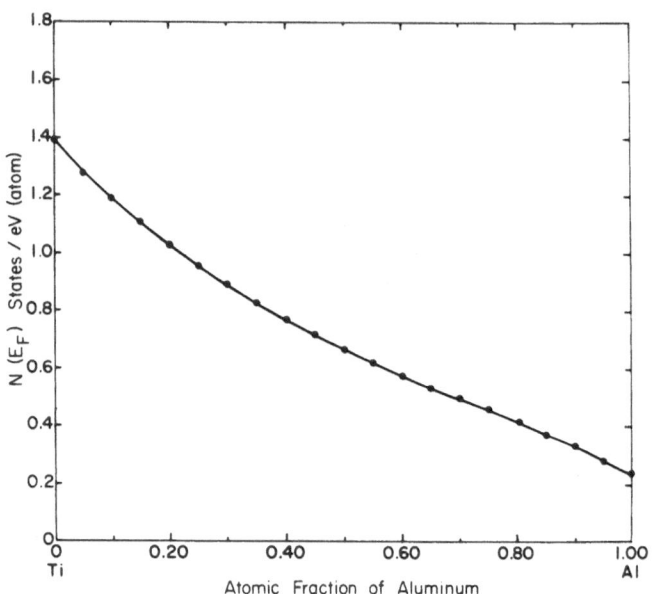

Fig. 17. Composition-dependent density of states at the Fermi
level E_F^a calculated for homogeneous Ti-Al alloys.

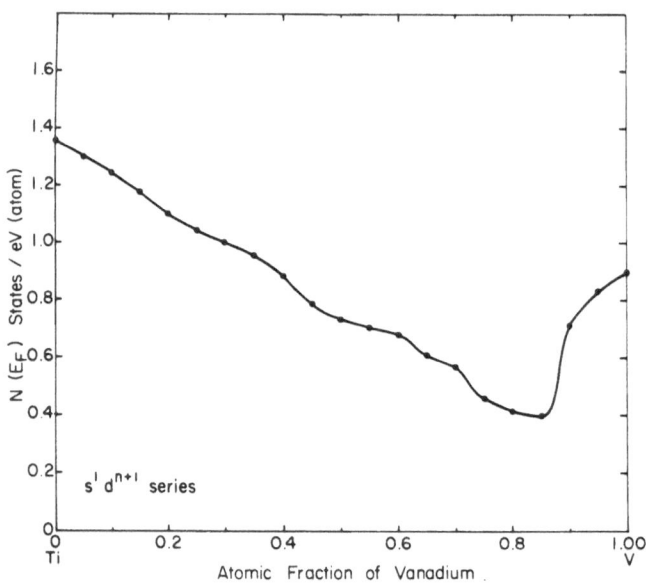

Fig. 18. Similar plot of $N(E_F^a)$ for Ti-V alloys.

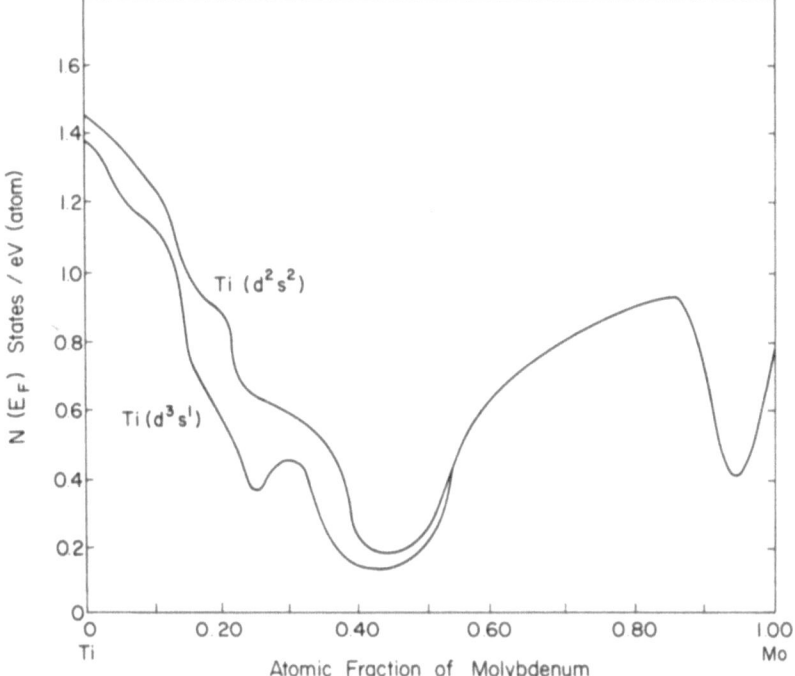

Fig. 19. Similar plot of $N(E_F^a)$ for Ti–Mo alloys. Separate curves are given for the configurations d^2s^2 and d^3s assumed for bcc-Ti by Waber and Snow [14].

with slowly changing slopes, as one saw for Al additions. The reason for the rapid variation of $N(E_F^a)$ with composition arises because the individual $N(E)$ curves have considerable amounts of structure. One even more striking $N(E_F^a)$ curve is presented for Zr and Mo in Fig. 20.

Before closing this topic, another illustration of the power of the LCDS model might be made. Calculations were done for the bcc and fcc forms of several pure metals. The calculated density-of-states curves are compared in Fig. 21. In the case of Fe, calculations by Koskimaki and Waber [61] showed that the α or bcc form

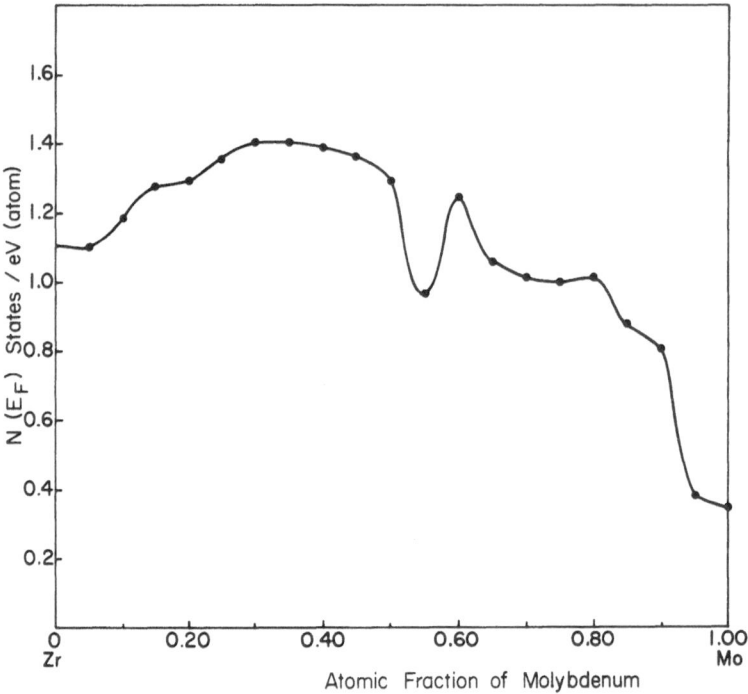

Fig. 20. Calculated density of states at the Fermi level for
         Zr-Mo alloys .

of Fe is more stable at room temperature than austenite. This was
established by integrating $E \cdot N(E)$ over the occupied states. The
transition energy occurred with the right sign, but was slightly
too large [61].

     Koskimaki and Waber investigated [18] two typical alloy addi-
tions to Fe using the LCDS model, namely, Cr and Mn. The total
alloy energies are shown for the two cases. The first, i.e., Fig.
22, shows that Mn, a well known austenite former, will stabilize
the γ phase at 0 K. The crossover point is surprisingly close to
the value determined from the Fe-Mn phase diagram. In contrast, as
Fig. 23 shows, no concentration of Cr will stabilize the γ phase
at 0 K.

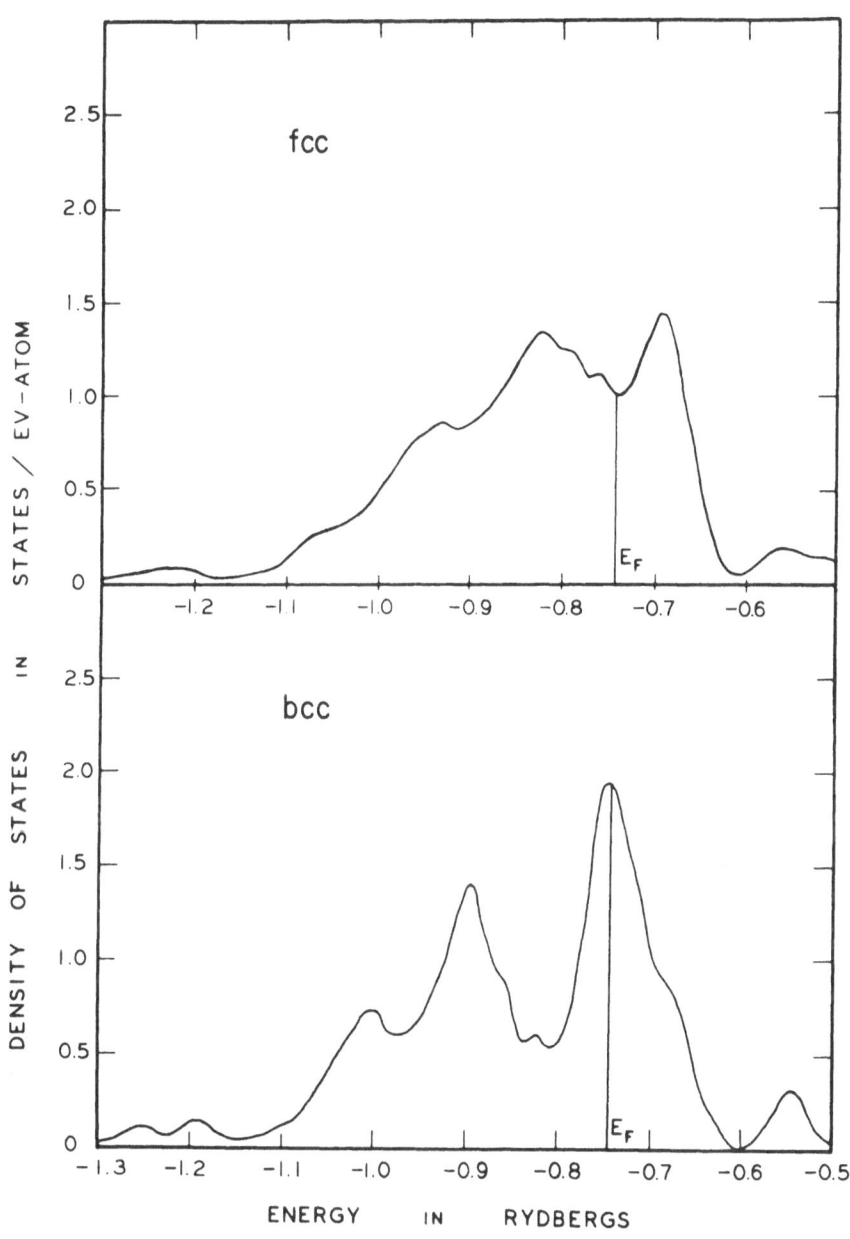

Fig. 21.   Comparison of the density of states curves for fcc and bcc
Fe in a paramagnetic state.   Note the difference in height and
slope at E_F (after Snow and Waber [14]).

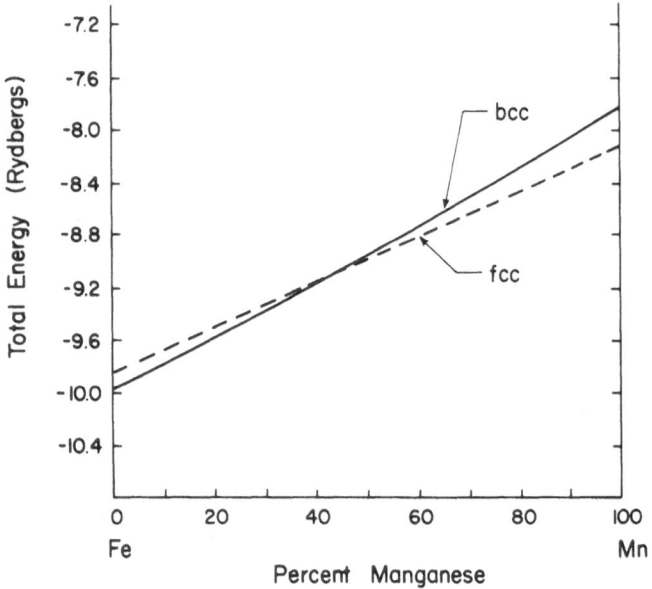

Fig. 22. Total energy of occupied electron states at 0 K as a func-
tion of composition using bands calculated for both pure phases with
Fe $(3d^64s^2)$ and Mn $(3d^54s^2)$. The transformation from fcc to bcc as
Mn is added is indicated.

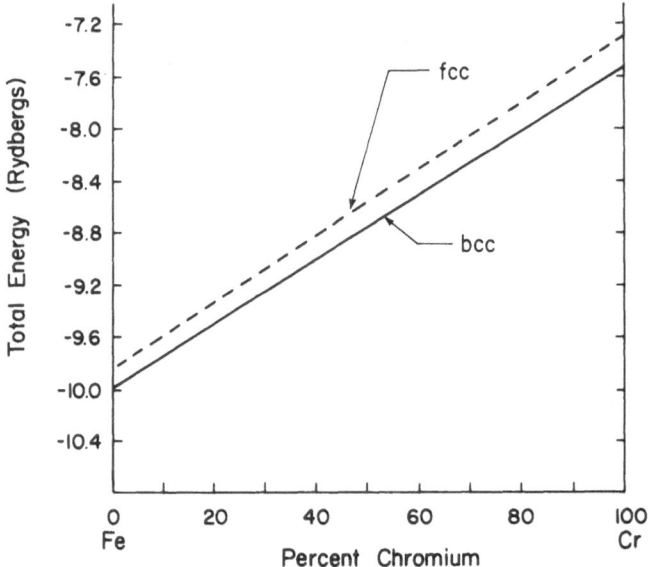

Fig. 23. A similar plot which shows that no Cr addition is sufficient to stabilize the austenite at 0 K. Based on data calculated
by Snow and Waber [14] using Cr with the $3d^44s^2$ input configuration.

A logical extension of this work would be to study the effect
of temperature on these two kinds of alloys. One defines formally
the energy $E_F(T)$, the Fermi level for any temperature, as the energy
at which the probability of a given energy state being occupied is
one-half--this is more general than the usual definition which only
pertains to 0 K. Specifically, one would multiply the alloy N(E)
curve by the Fermi distribution function

$$f(E,T,E_F) = \left[1 + \exp \left(\frac{E-E_F(T)}{KT} \right)^{-1} \right] \, , \qquad (23)$$

to obtain the repopulation of the electronic levels in an alloy.
With good luck, something approaching a phase diagram would be obtained.

It is worthwhile to emphasize here that no thermodynamic facts
have been built into the investigation. The atomic number and the
interatomic distance are the only pieces of information fed into
this to obtain individual density of states curves; and the raw N(E)
curves in Fig. 23 were then combined to get the results just presented. Incidentally, the metals were "mixed" in the paramagnetic
form and so the α/β Curie transformation was not used to bring about
the change in stability of the α and γ phases [62]. Many refine-
ments such as a better treatment of exchange, and the separate
handling of spin-up and spin-down states, could be introduced.
These would probably lead to an improved agreement with experimental
results. It is, however, striking that the present simple treatments
outlined above do lead to useful predictions and increased understanding.

4. HELLMAN-FEYNMAN CALCULATIONS

The Hellman-Feynman theorem shows that the force F, acting on
the pth nucleus of a solid or molecule can be deduced (almost)
classically from the total electronic charge density, even though
the latter is obtained from a quantum mechanical calculation. More
specifically, the x component of F can be written as

$$F_{px} = Z_p \left[\sum_q \frac{Z_q}{(R_p-R_q)^2} \left(\frac{\partial (R_p-R_q)}{\partial x_p} \right) - \int \left(\frac{\rho(r) \, x_p}{r_p^3} \right) d^3r \right] \, , \qquad (24)$$

where Z_p is the charge on that nucleus, x_p is the x-component of
r_p, and $\rho(r)$ is the total one-electron density such as is found in
a self-consistent calculation. That is, the total force on the
nucleus is the electrostatic force on it due to other nuclei plus
that due to the detailed distribution of surrounding electrons.

While this formulation is the adiabatic one, there have been dynamic treatments such as

$$M_p \; \ddot{x}_p \; = \; \left(\frac{\partial W_N}{\partial x_p}\right) - \int \rho(r,t) \left(\frac{\partial V}{\partial x_p}\right) d^3 r \quad, \tag{25}$$

where W_N is the internuclear coulombic force and $\partial V/\partial x_p$ is the instantaneous force acting on the nucleus due to varying charge density $\rho(r,t)$.

Deb [63] has made a very up-to-date review of the force concept deduced from the Hellman-Feynman theory. Slater [64] has re-derived the theorem based on the multiple-scattering Xα method that he has used for deriving molecular wave functions. The promise of this Hellman-Feynman theory has been less than expected and the difficulty has been attributed to the poor quality of molecular wave functions which have been available.

This theory looks particularly useful for obtaining the relative movement and rearrangement of atoms at a free or an internal interface. It also appears to have a ready application to absorption of gaseous species on a metal.

Slater [64] pointed out that the equilibrium spacing of the surface nuclei will increase if there is an excess charge density on the side away from bulk; it will pull the nuclei out. In the contrary situation, it will decrease the interatomic distance. The vector sum of forces arising from the non-spherical electron distribution will give rise to a surface pressure.

In 1969, Wannier et al. [65] calculated the force for Li atoms at the surface. They found a large positive internal pressure which would have led to considerable dilation. After pointing out some errors in Wannier's treatment, Kleinman [66] showed that the forces on individual atoms will vanish if one assumes a slight increase of surface charge by 0.049e and with a slight outward movement of the nuclei in the surface plane by only $0.00395a_0$. Without this dilation, Kleinman found a very substantial negative internal pressure localized near the surface.

5. DISCUSSION

Using an ionic model, Gilman [67] was able to show that the elastic compressibility modulus E of a series of AB compounds was proportional to the inverse 4th power of r_{AB}. When applied to metals, it was found that E depended on $r^{-\nu}$ where ν approached 5.

The compressibility is properly obtained in the following manner:

$$B = \frac{1}{K} = V \left(\frac{dR_s}{dV}\right)^2 \frac{d^2E_T}{dR_s^2} = \frac{1}{12\pi R_s}\left(\frac{d^2E_T}{dR_s^2}\right) \quad . \tag{26}$$

As an alternative procedure, one might consider calculating dP/dR_s from Eqn. (12) which would lead to a term in dV/dR_s. However, this equation is already another form of

$$P = -N \left(\frac{dR_s}{dV}\right)\left(\frac{dE_T}{dR_s}\right) \quad . \tag{27}$$

In either form, one would obtain an expression involving R_s^{-1} before the second derivative, but the numerical factor would be significantly different.

One can readily compute the second derivative of E_T with respect to R_s. This is

$$\frac{d^2E_T}{dR_s^2} = \frac{4[3NR_o^2 + 6\ell(\ell+1)]}{R_s^5} + \frac{3[6\ell(\ell+1) - 4.42\,a]}{R_s^4}$$

$$+ \frac{[3N - 1.2N - 0.916\,a]}{R_s^2} + \frac{4b}{\rho_{BM}^2}\exp\left(-\frac{S_{ij}}{\rho_{BM}}\right) \quad , \tag{28}$$

where a involves $\sqrt[3]{N}$.

Thus, B can be written as

$$B = \frac{-12NR_o^2}{12\pi R^6}\,\frac{24\ell(\ell+1)}{12\pi R_s^6} + \frac{18\ell(\ell+1) - 13.26a}{12\pi R_s^5} + \dots \tag{29}$$

Since both R_s and R_o are greater than unity, the lead terms become

$$B = \frac{N}{\pi}\left(\frac{R_o^2}{R_s^6}\right) + \frac{3\ell(\ell+1)}{2\pi R_s^5} \quad . \tag{30}$$

This deduction can be checked by using experimental values of B. The slope from a log-log plot of the experimental bulk moduli

versus R_S for various transition metals and metals with differing
valences and crystal structures was found to be larger than 6.
Subsequent plots given by Gilman [68] show that the slope is larger
than 5 for transition metals. The bulk moduli estimated by means
of the relation $B = C^2/V_0$ gave a relatively smooth downward trend
when plotted against R_S; the slope was more than 5. This is merely
a further indication of the type of success which the volume-
dependent FR model can be expected to yield.

One purpose of this paper is to emphasize how directly many
of the properties of metals and alloys are tied to the pressure of
the electronic gas which comes from the electrostatic repulsion
between the electrons. Equilibrium distances, heats of formation,
as well as compressibilities can readily be computed using slight
modifications of the Fröhlich-Raimes model.

In the FR model, the kinetic energy term is computed as though
the electrons were represented by a parabolic band, i.e., by a
structureless N(E) curve. But we know that such a description is
inadequate per se. To offset this deficiency, we have incorporated
a discussion of the use of the richly structured N(E) curves for
actual metals.

Let us turn to the LCDS model which is structure sensitive.
The volume-dependent effects are not included here, but they could
readily be incorporated when sufficient data on the variation of
the $E(\underset{\sim}{k})$ curves as a function of R_S become available.

One successful aspect of the LCDS model is that it offers a
simple logical reason for the deviations from the idealized Slater-
Pauling curve of unpaired spins N(E). For differing atomic numbers,
the N(E) curve may be very different in width and variable in location with respect to vacuum. The number of paired spins at E_F is
thus not simply related to the density of states curve of either
constituent since it depends on the response of the two partial spin-
up and two spin-down bands in a binary alloy to the internal magnetic
field βH. Even in a pure metal, $N(E_F\uparrow)$ is not equal to $N(E_F\downarrow)$, and
hence, transfer of electrons from one partial band to another occurs
and leads to an unbalance in the pairing of spins. If the two constituents are elements of very different atomic number, the effect
of a magnetic field on each N(E) curve may be quite different, while
the effect of solute additions may result in recognizable trends in
Slater-Pauling curves. There is no reason to anticipate that the
heights and the widths of the individual alloy curves could be superimposed when plotted against (e/a) values. In contrast, the Rigid
Band model assumes that the Fermi level and hence, the $N(E + \beta H)$
and $N(E - \beta H)$ values at E_F are readily calculated from (e/a) ratio
without knowing partial spin bands of the constituents.

Hummel et al. [69], using differential reflectivity, observed that the absorption edge (or change in the imaginary component ε_2 of the dielectric constant) associated with the d-bands of Cu, did not move to lower energy with Ni additions. The energy separation between the peak energy did not move. They also measured the reflectivity of Cu-Zn alloys and noted that the energy separation between the d-bands and E_F does increase with Zn content. This they rationalize by making the energy of the state L_1 the same in pure Cu as in α-brass. Thus, the energy level E_F should increase with C_A due to the increase in the mean number of valence electrons. These observations on Cu-Ni and Cu-Zn alloys are not inconsistent with the LCDS model.

Other theoretical predictions for Cu-Ni alloys have been made by Lang and Ehrenreich [70] and by Stocks, Williams, and Faulkner [71]. The former group predicts that $N(E_F^a)$ should decrease linearly with C_{Cu}, whereas the Oak Ridge group which used the CPA approach predict a more rapid initial decrease followed by a leveling off above 60% Cu. It is interesting to note that Huffner et al. [72] point out that their experimental density of states for the alloys "... (could) to a very good approximation, be made up by superimposing those of Ni and Cu."

To date, no synthesis of these distinct approaches has been attempted. Pragmatically, they are two limiting cases of the more complex true situation. While my colleagues and I are working on such a synthesis, it is hoped that this review will stimulate some readers to undertake the overall problem, and make a significant contribution to the science of metals.

ACKNOWLEDGMENT

The author would like to thank Dr. Allen C. Larson (Los Alamos Scientific Laboratory), who was of considerable help when this idea was first investigated, and Mr. F. W. Schonfeld and Prof. K. Gschneidner, who offered continuing encouragement over the intervening years. Because of these three individuals, a summary has at last been made available. The LCDS model arose from discussions with John Wood and A. C. Switendick. Exploration of the model can be attributed to the intensive work of David Koskimaki while he was a graduate student in the Materials Science Department of Northwestern University. To these individuals in particular, and to numerous colleagues who have offered guidance and criticism, the author acknowledges his indebtedness and apologizes for not citing many of them who have escaped his recollection. This research was supported in part by the National Aeronautics and Space Administration and by the U.S. Atomic Energy Commission. The work was initiated while the author was at Los Alamos Scientific Laboratory.

REFERENCES

1. BROOKS, H., Nuovo Cimento, $\underline{7}$ Supplement 2, 166-244 (1958).
2. WIGNER, E.P. and SEITZ, F., Solid State Physics, $\underline{1}$ 97-126 (1955).
3. FRÖHLICH, H., Proc. Roy. Soc., $\underline{A158}$ 97-110 (1937).
4. RAIMES, S., Phil. Mag., ser. 7, $\underline{43}$ 327-337 (1952).
5. RAIMES, S., Proc. Phys. Soc., $\underline{A66}$ 949-950 (1953).
6. BERNSTEIN, B.T., Iowa State University, thesis, 1959; Univ. Microfilms, Inc., Ann Arbor, Mich., No. Mic. 60-568, p 120.
7. LARSON, A.C. and WABER, J. T., Bull. Am. Phys. Soc. $\underline{8}$ 10 (1963).
8. LARSON, A.C., WABER, J.T. and SMITH, J.F., Los Alamos Scientific Lab and Iowa State Univ., unpublished research, 1964.
9. JONES, H., Proc. Roy. Soc., $\underline{A144}$ 225 (1934); $\underline{A147}$ 396 (1934); Proc. Phys. Soc., $\underline{A49}$ 250 (1937); Phil. Mag., $\underline{43}$ 105 (1952).
10. HUME-ROTHERY, W., SMALLMAN, R. and HAWORTH, C.W., Structure of Alloys, Institute of Metals, London, 1969.
11. SLATER, J.C., Phys. Rev. $\underline{55}$ 675 (1936).
12. An example of the Slater-Pauling curve is given by CRANGLE, P., Electronic Structure and Alloy Chemistry of the Transition Elements, BECK, P., ed., Interscience, New York, 1963, p 57.
13. RAIMES, S., J. Phys. and Rad. $\underline{23}$ 639 (1962).
14. SNOW, E.C. and WABER, J.T., Acta Met. $\underline{17}$ 64 (1969).
15. MANNING, M.F., Phys. Rev. $\underline{63}$ 190 (1943), and GREENE, J.B. and MANNING, M.F., Phys. Rev. $\underline{63}$ 203 (1943).
16. CHODOROW, M., Phys. Rev. $\underline{55}$ 675 (1939). See also KRUTTER, H.M., Phys. Rev. $\underline{48}$ 664 (1935) and FUCHS, K., Proc. Roy. Soc. $\underline{A151}$ 585 (1935).
17. WABER, J.T., see discussions in Electronic Density of States, BENNETT, L.H., ed., Spec. Publ. 323, National Bureau of Standards, 1971, pp 221, 323, 817.
18. KOSKIMAKI, D. and WABER, J.T., Northwestern Univ., unpublished research, 1971.
19. COLLINGS, E.W. and GEGEL, H., paper in this volume.
20. STERN, E.A., Phys. Rev. $\underline{B5}$ 368 (1972).
21. TRICKEY, S.B., AVERILL, F.W. and GREENE, F. R., Jr., Phys. Letters, $\underline{9}$ 385 (1972).
22. PERROT, F., Theses d'etat, Paris, 1971.
23. SNOW, E.C., "Total Energy as a Function of Lattice Parameter for Copper via the Self-Consistent APW Method", submitted to Phys. Rev. Also, "Compressibility of Copper via the Self-Consistent APW Method", submitted to Solid State Comm.
24. We have followed the traditional Wigner-Seitz boundary condition. However, the correct one is ℓ-dependent and the gradient vanishes for even values of ℓ; but it is necessary for the wave function to vanish at R_s for ℓ-odd, to insure the proper matching of wave functions in contiguous cells.
25. Even though the right-hand side of Eqn. (4a) may become negative, the term $\beta R_0 \cot \beta R_0$ remains real since the $\sin(i\chi) = i \sin \chi$ and $\cos(i\chi) = \cos \chi$. The logarithmic derivative in Eqn. (6), however, becomes a complex number unless A is set equal to zero.

26. SLATER, J.C., Phys. Rev. 81 385 (1951); Adv. Quant. Chem. 6 1 (1972). SCHWARZ, K., Phys. Rev. B5 2466 (1972). SLATER, J.C. and JOHNSON, K.H., Phys. Rev. B5 844 (1972). HEDIN, L. and LUNDQUIST, S., J. Phys. Colloq. C3 C3-7 (1972).

27. LIBERMAN, D., Phys. Rev. B3 2081 (1971).

28. PINES, D. and NOZIÉRES, P., Phys. Rev. 111 442 (1958).

29. GSCHNEIDNER, K., Rare Earth Alloys, Van Nostrand, New York, 1961, p 25.

30. TRULSON, O.C., HUDSON, D.E. and SPEDDING, F.H., Jr., Jour. Chem. Phys. 35 1018 (1961).

31. BRIDGMAN, P., The Physics of High Pressure, Strangeways Press, London, 1952, pp 160-161.

32. PEARSON, G.L. and BRATTAIN, W.H., Proc. Inst. Radio Eng. 43 1794-1806 (1955).

33. TEATUM, G., GSCHNEIDNER, K. and WABER, J., Los Alamos Scientific Laboratory Document LA 2345, June 1960, 235 pp, revised 1970.

34. MOORE-SITTERLY, C.E., National Bureau of Standards Circular 467, vols. I, II, and III, 1949.

35. FINKELNBURG, W. and HUMBACH, W., Naturw., 42 35 (1955). See also Phys. Rev. 77 303-304 (1950).

36. KLINKENBERG, P.F.A., VAN KLEEF, A.M. and NOORMAN, P.E., Physica, 27 151 (1961).

37. RAIMES, S., Imperial College, New York, unpublished research, 1951.

38. WABER, J.T. and LARSON, A.C., Rare Earth Research II, Gordon & Breach, New York, 1964, pp 351-383.

39. BROOKS, H., Trans. TMS-AIME, 227 546 (1963).

40. KUHN, T.S., Phys. Rev. 79 515 (1950).

41. McQUEEN, R.G. and MARSH, S.P. Jour. Appl. Phys. 31 1253 (1960). The Hugoniot is a pressure-volume based curve which allows for significant temperature rise, and is based on the Hugoniot-Rankine Equations. The curve marked 'isentrope' allows for the reduction of temperature to 0 K isentropically.

42. HAFEMEISTER, D. and ZAHRT, J., Journ. Chem. Phys. 47 1629 (1967). See also HAFEMEISTER, D., Jour. Chem. Phys. 43 789 (1965).

43. PEARSON, W.B., A Handbook of Lattice Spacings and Structures of Metals and Alloys, Pergamon Press, New York, 1958, pp 876-878.

44. WABER, J.T., HARRIS, I.R. and RAYNOR, G.V., Trans. Amer. Inst. Met. Eng. 230 148 (1964).

45. FRIEDEL, J., Phil. Mag. 46, ser. 7, 514 (1955).

46. GSCHNEIDNER, K. and VINEYARD, G.H., Jour. Appl. Phys. 33 3444 (1962).

47. HARRIS, I.R. and RAYNOR, G.V., Univ. of Birmingham, private communication, 1961.

48. DAVISON, J.E. and SMITH, J.F., Trans. TMS-AIME, 242 2045 (1968).

49. SHANNETTI, G.W. and SMITH, J.F., Jour. Appl. Phys. 42 2799 (1971). SAMER, A. and SMITH, J.F., Jour. Appl. Phys. 33 2283 (1962).

50. BOLSAITIS, P., Met. Trans. 4 2395-2398 (1973).

51. CHIARODO, R., GREEN, J., SPAIN, I.L. and BOLSAITIS, P., Jour. Phys. Chem. Solids, 33 1905-1914 (1972).

52. HSIEH, K. and BOLSAITIS, P., Jour. Phys. Chem. Solids, 33 1838-1842 (1972).

53. SANDERSON, R.T., Chemical Periodicity, Reinhold Publishing Company, New York, 1960, p 25.

54. JOHNSON, K.H. in Energy Bands in Metals and Alloys, BENNETT, L.H. and WABER, J.T., eds., Gordon & Breach, New York, 1968, p 108.

55. JOHNSON, K.H. and CONNOLLY, J.W.D., Int. Jour. Quant. Chem. III 813 (1970).

56. SOVEN, P., Phys. Rev. 156 809 (1967). See also Fig. 1 of article by SOVEN in Energy Bands in Metals and Alloys, BENNETT, L.H. and WABER, J.T., eds., Gordon & Breach, New York, 1968, p 146.

57. For example, see EASTMAN, D.E., CASHMAN, J.K. and SWITENDICK, A.C., IBM Company Report RC 5368, 1971. EASTMAN, D.E., Proc. Conf. on Electron Spectroscopy, Asilomar, 1971. For pure metals, see EASTMAN, D.E., Jour. Appl. Phys., 40 1387 (1969) and JANAK, J.F., EASTMAN, D.E. and WILLIAMS, A.R., Solid State Comm. 8 271 (1970).

58. SEIB, D.H. and SPICER, W.E., Phys. Rev. Letters, 20 1441 (1968); also, Phys. Rev. B2 1676 and 1694 (1970); SPICER, W.E., Phys. Rev. 154 385 (1967); BERGLAND, C.N. and SPICER, W.E., Phys. Rev. 136 A1030 and A1044 (1964).

59. CLIFT, J., CURRY, C. and THOMPSON, B.J., Phil. Mag. 8 592 (1963).

60. BENNETT, L.H. and WILLENS, R., eds., Charge Transfer/Electronic Structure of Alloys, Met. Soc. of the Am. Inst. of Mining, Met. and Pet. Eng., New York, 1974.

61. KOSKIMAKI, D. and WABER, J.T. in Electronic Density of States, BENNETT, L.H., ed., Spec. Publ., National Bureau of Standards, 1971, p 741.

62. Investigation of temperature effects in repopulating various states was investigated; just using the quantum statistics, it was shown that γ-Fe becomes more stable with increasing temperature, but Fe converts back to the bcc δ-Fe at sufficiently high temperatures, due to electron-phonon interaction effects.

63. DEB, B.M., Rev. Mod. Phys. 45 22-43 (1973).

64. SLATER, J.C., Jour. Chem. Phys. 57 2389 (1973).

65. WANNIER, G.H., MISNER, C. and SCHAY, G., Phys. Rev. 185 983 (1969).

66. KLEINMAN, L. Phys. Rev. B3 3083 (1971).

67. GILMAN, J.J., National Bureau of Standards Monograph No. 59, 1963, pp 79-102.

68. GILMAN, J.J., in Progress in Ceramic Science, BURKE, J.E., ed., Pergamon Press, New York, 1961, pp 146-199.

69. HUMMEL, R.E., HOLBROOK, J.A. and ANDREWS, J.B., Surface Science, 37 717 (1973).

70. LANG, N.D. and EHRENREICH, H., Phys. Rev. 152 520 (1966).

71. STOCKS, G.M., WILLIAMS, R.W. and FAULKNER, J.S., Phys. Rev. Letters, $\underline{26}$ 354 (1971); see also Phys. Rev. $\underline{B4}$ 4390 (1971).
72. HUFFNER, G., WERTHEIM, G.K., COHEN, R.L. and WERNICK, J.H., Phys. Rev. Letters, $\underline{28}$ 438 (1972).

UNRESOLVED SEGREGATION EFFECTS IN SOLID

SOLUTION STRENGTHENING OF METALS

W. C. Leslie

Materials and Metallurgical Engineering Department
The University of Michigan
Ann Arbor, Michigan 48104

ABSTRACT

The evidence for grain boundary and free surface segregation of solutes is reviewed. The effects on macroscopic mechanical properties of grain boundary segregation of solutes in metals have been established in only a few Cu-base, Ni-base and Fe-base systems. In each instance, the method employed has been the determination of σ_o and k in the Hall-Petch relation between yield strength and grain size:

$$\sigma = \sigma_o + kd^{-n}.$$

Solute atoms can increase or decrease k, or have no effect. The effect of solute concentration on k is not always linear. Apparently, the effects of interdendritic segregation on macroscopic mechanical properties of metallic solid solutions have never been determined.

1. INTRODUCTION

There are certain difficulties which intrude in experimental tests of theories of solid solution effects in metals. The two most common, and the subjects of this paper, are solute segregation at grain boundaries in polycrystalline specimens and local concentrations of solutes arising from interdendritic segregation during freezing. Until quite recently these inhomogeneities were generally overlooked or deliberately ignored, and with good reason. Determination of the existence and the magnitude of their effects on

275

mechanical properties is difficult or tedious or both. However, if
we are to progress in our understanding of solid solutions, we must
now face these complications and devise means of dealing with them.

This brief paper is an attempt to summarize the current state
of knowledge of the effects of grain boundary and interdendritic
segregation of solutes on the macroscopic mechanical properties of
metallic solid solutions.

2.   EVIDENCE AND EFFECTS OF GRAIN BOUNDARY SEGREGATION

2.1   Grain Boundary Hardening

A commonly used method for detecting grain boundary segregation of solutes in metals is the determination of hardness changes
in the traverse of a grain boundary. The subject, reviewed some
years ago by Westbrook [1,2] continues to be an active field of
study [3-7]. The hardening which usually accompanies the segregation may extend over regions that are much wider than the grain
boundary. The total width affected can be about 50 μm [4], as
compared with a grain boundary width of about three atom diameters
[8]. The currently accepted explanation of this relatively wide
band of segregation is that it is a result of a vacancy-coupled
flow of solute atoms [9]. Vacancies tend to migrate to sinks, such
as free surfaces and grain boundaries, as a specimen is cooled from
high temperature. At these sinks the vacancies are annihilated,
leaving a local non-equilibrium concentration of solute atoms. It
follows that the extent of grain boundary segregation and hardening
is a function of the thermal history of the specimen. This type
of segregation can also occur at exterior surfaces [6,19] where it
will affect mechanical properties [10]. In this circumstance we
have essentially no knowledge of the nature or magnitude of such
effects.

The occurrence of this kind of non-equilibrium segregation
does not preclude equilibrium (Gibbsian) segregation at interfaces,
about which more will be said later.

2.2   Auger Spectroscopy

Direct evidence of grain boundary segregation has been obtained
by Auger electron spectroscopy, principally from fracture surfaces
in temper-embrittled steels [11-14]. This work was primarily concerned with segregation of P or Sb at prior austenite grain boundaries. All the evidence indicates that the segregation in these
instances is limited to a region within 10 Å of the boundary. This,
plus the fact that the amount of segregate seems to satisfy the

Gibbs adsorption isotherm, has led to the conclusion that equilibrium segregation is involved.

These results from Auger spectroscopy were at least partially
confirmed by Low and Smith [15]. They dissolved layers 100 to 200 Å
thick from the intergranular fracture surfaces of temper-embrittled
AISI 3340 steel. Neutron activation analysis of the residues of the
solutions showed that Sb was segregated to the prior austenite grain
boundaries.

The usefulness of Auger spectroscopy in the analysis of fracture surfaces is not limited to steels; Joshi and Stein [16] found
segregation of P at intergranular fracture surfaces of W, and were
able to relate the ductile-brittle fracture transition temperature
to the extent of grain boundary segregation. As in the steels, the
P in the W was concentrated in the first few atomic layers at the
grain boundaries.

Rellick and his coworkers [17] added 0.02% Te to an Fe-0.02%
C alloy. The Te segregated strongly to both austenite and ferrite
grain boundaries, causing severe embrittlement. The concentration
of Te at the ferrite grain boundaries after slow cooling reached
about 25 at.%. These boundaries had an unusual faceted structure.

Joshi and Stein [18] detected S segregation to grain boundaries
in commercial purity Type 304 stainless steel, after a high-temperature solution treatment. Holding at 850 or 600°C after the solution
treatment led to grain boundary segregation of Si and Ni and to grain
boundary depletion of Cr.

Ramasubramanian and Stein [19] have found that S and P segregate to ferrite grain boundaries at all temperatures, but that
segregation in austenite is negligible. The segregation in ferrite
is irreversible and does not seem to be an equilibrium phenomenon.

Auger spectroscopy is equally effective in measuring solute
segregation at free surfaces. Barnes, et al. [20] found that the
concentration of Mo at the surface of mill-processed 0.05 in. thick
Type 316 sheet was about 14% as compared with 1.95% in the bulk.
This high concentration was limited to a region within 10 Å of the
surface.

2.3 Field Ion Microscopy

In two recent instances, grain boundary segregation of solutes
has been detected by field ion microscopy. Smith and Smith [21]
found that in oxygen-treated W the O content at grain boundaries was

about 30 times higher than in the bulk at about 500 Å from the
boundary. Because of the grain boundary embrittlement caused by O,
this observation is not surprising. More unexpected was the nearly
complete segregation of Cr to grain boundaries in W, discovered by
Howell et al. [22]. The average concentration of Cr was 30 ppm, but
in the boundaries the concentration was in the range of 6 to 12%.
No short-range ordering or clustering was seen. No segregation
occurred at coherent twin boundaries.

2.4 Variation of k in the Hall-Petch Relation

It is now established that both equilibrium and non-equilibrium
segregation of solutes can occur at metallic grain boundaries. The
questions remaining to be answered are how and to what extent such
segregation affects the macroscopic mechanical properties of the
metal. There have been only a few attempts to determine the magnitude of such effects and these have all relied on determining change
of the slope, k, in the Hall-Petch relation,

$$\sigma = \sigma_o + kd^{-n},$$

where σ is the yield or flow stress and d is average grain diameter.
Armstrong [23] has reviewed much of the earlier testing of this relation. Despite considerable effort, there is still some uncertainty
regarding the value of the exponent, n. Morrison [24] tested the
relation in Fe and in carbon steels over the widest range of grain
size that has been employed, with d ranging from 1.6 to 400 μm.
For the lower yield stress, his results show that a $d^{-1/2}$ relation
gives a better fit to the data than either $d^{-1/3}$ or d^{-1}. On the
other hand, Hutchison and Pascoe [25], working with Cu-base alloys,
concluded that the best value for n was -1. Their range of grain
sizes was about 11 to 200 μm, a considerably smaller range than
Morrison's. Also, Hutchison and Pascoe [25] equated σ to the stress
at 1% strain. Although many of their alloys displayed a Lüders
strain, they chose 1% strain in preference to initial or lower
yield stress because of the higher effective strain rate during the
propagation of a Lüders front. However, in view of the very low
strain rate sensitivity for most fcc alloys, this does not seem a
sufficient reason to abandon the lower yield stress, or the stress
at a smaller offset, such as 0.2%.

Fujita and Tabata [26], working with high-purity, polycrystalline Al, found that the relation between lower yield stress and
grain size was best expressed by $d^{-1/2}$. In agreement with Morrison
[24] they also found that stresses at larger strains deviated from
the Hall-Petch relation.

The bulk of available evidence indicates that the relation between yield stress and grain size in most metals and alloys is satisfied by the equation:

$$\sigma_{ys} = \sigma_o + kd^{-1/2}.$$

The models proposed to account for this type of grain size dependence of yield stress have been summarized by Hirth [27]. Detailed dislocation models involving grain boundary pile-ups or grain boundary emission lose their attractiveness in the face of Grange's observation [28] that a $d^{-1/2}$ relation exists between the yield strength of steels and the prior austenite grain size. This held regardless of whether the structure tested was ferrite-pearlite, upper bainite, lower bainite or tempered martensite, or whether the structure was obtained by continuous cooling, isothermal transformation, or quenching and tempering. The austenite grain sizes ranged from 3 to 100 μm. The situation is complicated by the changes in carbide size and morphology that accompany the change in austenite grain size, but it is striking that all such changes can be encompassed in the $d^{-1/2}$ relation.

It appears that a more macroscopic approach to the problem of grain-size strengthening may be fruitful. An effort in this direction has been made by Kocks [29] and by Hirth [27], who assumed that each grain behaved like a cell with a hard crust and a soft interior. Unfortunately, they predicted values of n ranging from –1 to –2, which does not agree with most observations.

When a solute is added to a single-phase, high-purity metal, k in the Hall-Petch relation changes with the type and concentration of the added element. The effect on determination of solid solution strengthening is serious, as illustrated in Fig. 1, taken from Hutchison and Pascoe [25]. This shows the result of a change with grain size of relative strengthening by solutes.

There appear to have been only three systematic studies of this phenomenon. The first was by Floreen and Westbrook [5], who found that as increasing amounts of S were added to high-purity polycrystalline Ni the value of k first increased, went through a maximum, then decreased to less than the value for the unalloyed Ni.

The most extensive study was conducted by Hutchison and Pascoe [25], who alloyed Cu with Ag, Al, Au, Ga, Ge, In, Ni, Sb, Sn or Zn, each at several concentrations. They found that some solutes increased k whereas others decreased it, but in each instance k varied linearly with solute concentration.

In Fe-base alloys, the only study of the effect of substitutional solutes on the Hall-Petch slope is that reported by Morrison

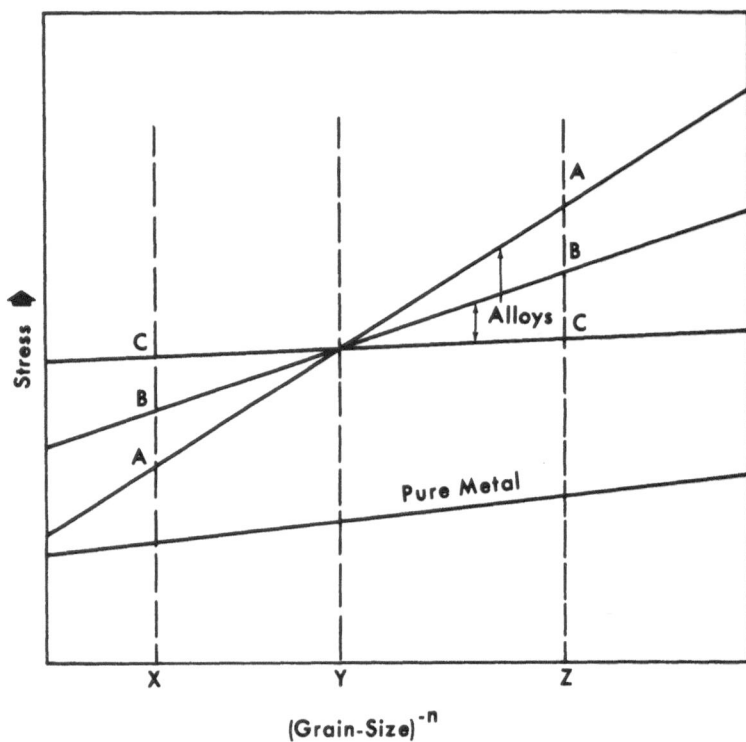

Fig. 1. Hypothetical relationships between stress and (grain-size)$^{-n}$, where n is a constant, for a pure metal and three alloys containing the same concentrations of solutes A, B and C. The order of strengthening due to these solutes is (a) ABC at grain-size X, (b) the same at Y, and (c) CBA at Z. (Hutchison and Pascoe [25])

and Leslie [30]. They found that when tested at room temperature, Ni and Si increased k, but Cr had no effect. Small concentrations of C or N can greatly increase k. Confirmation of the increase of k by Ni in solution in Fe is obtained from the paper by Jolley [31]. At 3.38 wt.% Ni, his value of k is the same as that of Morrison and Leslie for 3.08 wt.% Ni.

In view of the solid solution softening phenomenon in bcc alloys, it is interesting to note that k for Fe at -75°C is <u>greater</u> than k for Fe containing 1.5 or 3.0 at.% Si. As the test temperature is decreased below ambient, k for Fe increases markedly, but k for the Fe binary alloys remains unchanged [30].

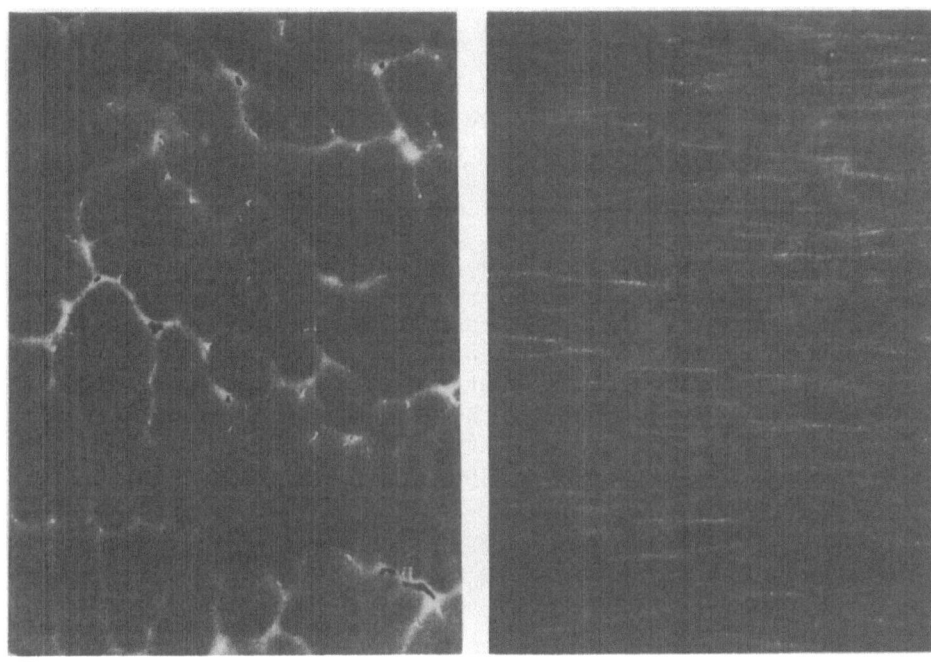

(A) As Cast   (B) As Rolled

Fig. 2. Manganese segregation in an Fe-1.5Mn-0.2C alloy cast into
a 3-inch diameter mold (a) and subsequently hot-rolled to 1/2-inch
plate (b).   [75X]

On the basis of the available evidence, we must conclude that
in any comprehensive study of solid solution effects in polycrystalline metals the effects of solutes on σ_0 and k in the Hall-Petch
relation must be determined [24]. In addition, at least for bcc
solvent metals, the changes in these parameters with temperature
must be considered.

3.   INTERDENDRITIC SEGREGATION

In studies of solid solution effects in metals, it is customary
to assume that the solute atoms are uniformly dispersed in the
matrix. This assumption is probably more often incorrect than not,
even if we ignore grain boundary segregation. Freezing of alloys
is frequently incongruent, resulting in interdendritic segregation
of the solute. An example of this is an Fe-Mn-C alloy shown in
Fig. 2, taken from unpublished work by Miller and Leslie [32]. The
two micrographs illustrate microsegregation of manganese (light
areas) revealed by microradiography. Cobalt K_α radiation was

used. Cr in Fe segregates in a similar manner; indeed, a microradio-graph of a ferritic stainless steel resembles a picture of a compos-ite material.

Once formed, such segregates are difficult to eliminate. Homo-genization is a standard problem in diffusion [33]; it can be hastened by mechanical working to reduce the segregate spacing followed by high-temperature annealing below the solidus.

We do not know to what extent solid solution effects are per-turbed by interdendritic segregation of solutes; as a minimum, directionality will be present in specimens. The subject should receive some attention, to establish, at least, the approximate magnitude of the maximum effect that can be expected. The solute element chosen should be one which produces a marked strengthening of the solvent. In other systems, such as Cr in Fe, the solute produces such minor strengthening that segregation effects can be ignored [34].

4. CONCLUSIONS

(a) Equilibrium and non-equilibrium segregation of solute atoms to grain boundaries and external surfaces adds an element of un-certainty to the determination of solid solution effects in metals.

(b) The magnitude of interface effects varies with solute species, concentration and thermal history.

(c) The Hall-Petch relation between yield strength and grain size affords a method for determining the macroscopic effects of grain boundary segregation, through determination of σ_0 and k.

(d) The macroscopic effects of interdendritic segregation of sol-utes in metals remain to be determined.

REFERENCES

1. WESTBROOK, J.H., Met. Rev. $\underline{9}$ 415 (1964).
2. WESTBROOK, J.H., Interfaces Conference, Melbourne, Australia, p 283, Butterworths, 1969.
3. BRAUNOVIĆ, M., HAWORTH, C.W. and WEINER, R.T., Met. Sci. J. $\underline{2}$ 67 (1968).
4. BRAUNOVIĆ, M. and HAWORTH, C.W., Proc. 2nd Int. Conf., Strength of Metals and Alloys, p 311, ASM, Cleveland, OH, 1970; Met. Sci. J. $\underline{4}$ 85 (1970).
5. FLOREEN, S. and WESTBROOK, J.H., Acta Met. $\underline{17}$ 1175 (1969)

6. AUST, K.T. and WESTBROOK, J.H., Acta Met. 19 521 (1971).
7. AUST, K.T. and CHALMERS, B., Met. Trans. 1 1095 (1970).
8. MORGAN, R. and RALPH, B., Acta Met. 15 341 (1967).
9. HANNEMAN, R.E. and ANTHONY, T.R., Acta Met. 17 1133 (1969).
10. LATANISION, R.M., SEDRIKS, A.J. and WESTWOOD, A.R.C., RIAS Tech.
    Rep. 71-06C, Feb. 1971.
11. MARCUS, H.L. and PALMBERG, P.W., Trans. AIME 245 1664 (1969).
12. STEIN, D.F., JOSHI, A. and LAFORCE, R.P., Trans. ASM 62 776
    (1969).
13. PALMBERG, P.W. and MARCUS, H.L., Trans. ASM 62 1016 (1969).
14. VISWANATHAN, R., Met. Trans. 2 809 (1971).
15. LOW, J.R., Jr. and SMITH, C.L., Carnegie-Mellon Univ. Report
    No. 031-727-3, May, 1971.
16. JOSHI, A. and STEIN, D.F., Met. Trans. 1 2543 (1970).
17. RELLICK, J.R., McMAHON, C.J., Jr., MARCUS, H.L. and
    PALMBERG, P.W., Met. Trans. 2 1492 (1971).
18. JOSHI, A. and STEIN, D.F., Corrosion, 28 321 (1972).
19. RAMASUBRAMANIAN, P.V. and STEIN, D.F., Met. Trans. 4 1735 (1973).
20. BARNES, G.J., ALDAG, A.W. and JERNER, R.C., J. Electrochem.
    Soc. 119 684 (1972).
21. SMITH, D.A. and SMITH, G.D.W., Proc. 3rd Int. Conf. Strength
    of Metals and Alloys, 1 144 (1973).
22. HOWELL, P.R., FLEET, D.E., PAGE, T.F. and RALPH, B., Proc. 3rd
    Int. Conf. Strength of Metals and Alloys, 1 149 (1973).
23. ARMSTRONG, R.W., Met. Trans. 1 1169 (1970).
24. MORRISON, W.B., Trans. ASM, 59 824 (1966).
25. HUTCHISON, M.M. and PASCOE, R.T., J. Aust. Inst. Metals, 14
    306 (1969); Aust. Def. Sci. Ser., Aero. Res. Lab., Metallurgy
    Report 85, Sept. 1971.
26. FUJITTA, H. and TABATA, T., Acta Met. 21 355 (1973).
27. HIRTH, J.P., Met. Trans. 3 3047 (1972).
28. GRANGE, R.A., Trans. ASM, 59 26 (1966).
29. KOCKS, U.F., Met. Trans. 1 1121 (1970).
30. MORRISON, W.B. and LESLIE, W.C., Met. Trans. 4 379 (1973).
31. JOLLEY, W., Trans. TMS-AIME 242 306 (1968).
32. MILLER, R.L. and LESLIE, W.C., unpublished work, E.C. Bain Lab.,
    U.S. Steel Corp., 1967.
33. SHEWMON, P.G., in Physical Metallurgy, CAHN, R.W., Ed., p 370,
    North Holland, Amsterdam, 1965.
34. LESLIE, W.C., Met. Trans. 3 5 (1972).

CONTRIBUTORS

BEELER, J.R., Jr.
  Nuclear Engineering Department
  North Carolina State University
  Raleigh, North Carolina 27607

BENEDEK, R.
  Department of Materials Science
    and Engineering
  Cornell University
  Ithaca, New York 14850

COLLINGS, E.W.
  Battelle-Columbus Laboratories
  505 King Avenue
  Columbus, Ohio 43201

CONRAD, H.
  Metallurgical Engineering and
    Materials Science Department
  University of Kentucky
  Lexington, Kentucky 40506

DE MEESTER, B.
  Metallurgical Engineering and
    Materials Science Department
  University of Kentucky
  Lexington, Kentucky 40506

DÖNER, M.
  Metallurgical Engineering and
    Materials Science Department
  University of Kentucky
  Lexington, Kentucky 40506

FISHER, E.S.
  Materials Science Division
  Argonne National Laboratory
  Argonne, Illinois 60439

GEGEL, H.L.
  Air Force Materials Laboratory
  Wright-Patterson Air Force Base
  Ohio 45433

GEHLEN, P.C.
  Battelle-Columbus Laboratories
  505 King Avenue
  Columbus, Ohio 43201

HIRTH, J.P.
  Department of Metallurgical
    Engineering
  The Ohio State University
  Columbus, Ohio 43210

HOAGLAND, R.G.
  Battelle–Columbus Laboratories
  505 King Avenue
  Columbus, Ohio 43201

HO, P.S.
  IBM Thomas J. Watson Research
    Center
  Yorktown Heights, New York
  10598

LESLIE, W.C.
  Materials and Metallurgical
    Engineering Department
  The University of Michigan
  Ann Arbor, Michigan 48104

OKAZAKI, K.
  Metallurgical Engineering and
    Materials Science Department
  University of Kentucky
  Lexington, Kentucky 40506

STERN, E.A.
  Department of Physics
  University of Washington
  Seattle, Washington 98195

TYSON, W.R.
  Physical Metallurgy Research
    Laboratories
  Department of Energy, Mines
    and Resources
  Ottawa, Ontario
  Canada

WABER, J.T.
  Department of Materials Science
  Northwestern University
  Evanston, Illinois 60201

AUTHOR INDEX

MATERIALS INDEX

SUBJECT INDEX

A

α-Fe, impurity effects, 117-144
α-stabilizing elements, 152
ACOUSTIC WAVE ATTENUATION, 201
   influence of phase changes or
      interstitial gases, 205
ACTIVATION DISTANCE, 23-26, 36,
   48, 50, 56, 60, 62
ACTIVATION ENERGY
   for impurity diffusion, 93ff
ACTIVATION LENGTH, 50, 56
ACTIVATION VOLUME, 19, 23
ANISOTROPIC ELASTIC THEORY with
   reference to dislocation
   interaction, 102
ANISOTROPY RATIO--see Elastic
   Modulus Anisotropy Ratio
ARRHENIUS-TYPE RATE EQUATION, 17
ASHCROFT "empty core" potential,
   82
ATOMIC bonding, 27, 40, 72, 73
ATOMIC configuration, 186, 187
ATOMIC diffusion, 79
ATOMIC interaction, local, 102
   use of Fermi-Dirac
      statistics, 102
   finite range of, 86
ATOMIC model of dislocation, 104

ATOMIC POTENTIAL
   difference, 147, 172
   amplitude and phase as function
      of model radius, 83
   Johnson, 104
   strength of, 148
AUGER SPECTROSCOPY, 276, 277
AUSTENITE GRAIN SIZE, 279
AVERAGE GROUP NUMBER, 155

B

B-sub-group metals, 149
BAND STRUCTURE ENERGY, 81
BARRIERS, dislocation, 51
BASIS VECTORS, 87, 90
BINDING ENERGY, 123-126, 128,
   138-143
   for carbon in α-Fe, 138-143
   for SIA and LIA complexes
      in α-Fe, 124
   for vacancy-LIA complexes
      in α-Fe, 128
   for vacancy-SIA complexes
      in α-Fe, 125
   impurity-vacancy, 80
   vacancy-impurity complexes,
      123-126